CONTENTS

Studying Reptiles and Amphibians
Snakebites

REPTILES

TURTLE

Usually four-legged (except for snakes and a few lizards); each foot has three to five clawed toes. Skin usually has horny scales, sometimes bony plates. Most lay eggs with a hard or leathery shell.

16–17

TURTLES Bony or leathery shell. Four limbs, short tail. Head can be withdrawn wholly or partly into shell. **18–43**

LIZARD

LIZARDS In the United States, mostly small, fast-moving land reptiles. Four-legged, covered with equal-sized horny, smooth, or beaded scales. Most lay eggs. **44–69**

SNAKES Long, flexible body, legless. Scales on belly usually larger than others. Skulls loose, mouth large. Lack ear openings. Some are egg-laying; some live-bearing. **70–113**

SNAKE

ALLIGATORS and CROCODILES Large, lizard-like. Skull forms long snout. Adapted to life in water in warm regions. **114–115**

AMPHIBIANS

Four- or, rarely, two-legged (except tadpoles). Smooth or warty skin, usually moist. No visible scales. Toes never clawed. Eggs are usually laid in jelly-like masses in water. **116–117**

ALLIGATOR

FROGS and TOADS Adults have much larger hind limbs than forelimbs; tadpoles limbless when young. Adults lack true tails. **118–136**

FROG

SALAMANDERS Most have four limbs, even the larvae. Limbs all about same size. Adults have tails. **137–153**

SALAMANDER

3

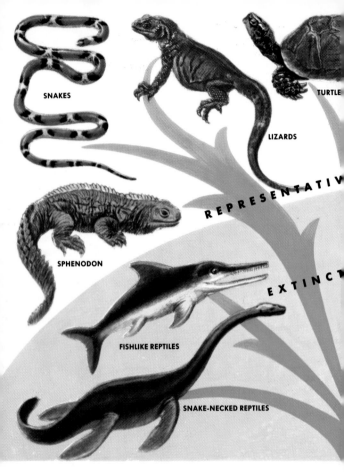

SNAKES

LIZARDS

TURTLE

REPRESENTATIV

SPHENODON

FISHLIKE REPTILES

EXTINCT

SNAKE-NECKED REPTILES

REPTILES evolved from amphibians nearly 250 million years ago. They slowly branched out and about 60 million years later dinosaurs and other reptiles became the

212 SPECIES IN FULL COLOR

REPTILES AND AMPHIBIANS

a Golden Guide® from St. Martin's Press

by Herbert S. Zim
and Hobart M. Smith

Revised by
Jonathan P. Latimer and
Karen Stray Nolting

with J. Whitfield Gibbons
University of Georgia's
Savannah River Ecology Laboratory

Illustrated by James Gordon Irving

St. Martin's Press 🅼 New York

FOREWORD

This book, along with other Golden Guides, has intro-
duced thousands of children and adults to the diversity of
the natural world. It illustrates the most familiar species
and gives concise information to help identify each one.
This revision reflects the latest information. We hope it
will continue to be an inviting primer on reptiles and amphibians for readers of all ages.

Many individuals and institutions have helped make
this guide a success. Foremost are Hobart M. Smith, who
co-authored the original edition, Herbert S. Zim, who
conceived Golden Guides and did so much to nurture their
growth, and James Gordon Irving, whose illustrations are
as useful today as they were when they were first painted.
Invaluable contributions were made to the first edition by
Charles M. Bogert, Bessie M. Hecht, James A. Oliver, Carl
F. Kauffeld. Roger and Isabelle Conant, Robert C. Miller,
Joseph R. Slevin, Earl S. Herald, L. M. Klauber, C. B.
Perkins, C. S. Shaw, Louis W. Ramsey, and William H.
Stickel. Philip and Dorothy Smith, Harold Kerster, Donald
Hoffmeister, Grace Crowe Irving, Rozella Smith, and So-
nia Bleeker Zim also provided helpful assistance.

Special thanks go to J. Whitfield Gibbons, Tony M.
Mills, and Brian S. Metts of the University of Georgia's
Savannah River Ecology Laboratory Herpetology Lab for
their contributions to this revision. We are also grateful to
David Challinor of the Smithsonian Institution for his
guidance.

J. P. L.

K. S. N.

ISBN 1-58238-131-3

BIRDS

CROCODILIANS

LIVING FORMS

TOOTHED BIRDS

FORMS

DINOSAURS

FLYING REPTILES

MAMMALS

MAMMALLIKE REPTILES

dominant lifeforms on the land. Some reptiles took to the air and to the seas. Reptiles of today are the descendants of such magnificent ancestors.

SPADEFOOTS and KIN

TOADS and KIN

SLIMY SALAMANDER and KIN

SURINAM, MEXICAN BURROWING TOADS and KIN

TRUE FROGS AND NARROWMOUTH TOADS and KIN

REPRESENTATIVE LIVING FORMS

TAILED AND MIDWIFE FROGS and KIN

EXTINCT

ERYOPS

REPTILES

SEYMOURIA

AMPHIBIANS had at least a 50-million-year head start on reptiles, but these first land animals never became completely independent of water. Their jelly-like eggs could not survive in air, so amphibians had to re-

NEWTS

MUDPUPPY

SIRENS

TIGER
SALAMANDER
and KIN

HELLBENDER

CAECILIANS

FORMS

LYSOROPHUS

DIPLOCAULUS

turn to swamps, ponds, or streams to breed. Ancestors of present-day frogs and salamanders flourished in the shallow prehistoric swamps. Many were clumsy giants. Their American descendants are found on pages 116–153.

DESERT IGUANA

REPTILES AND AMPHIBIANS

COMMON TOAD

There are many untruths told about reptiles and amphibians. Some are exaggerations and others are simply mistakes. For example, many people believe that snakes have slimy skins. If you actually touch a snake, you will find that its skin is dry. Many people are also afraid that a snake or lizard will attack them. Most reptiles and amphibians are shy and will escape if possible. Each year there are fewer than 15 deaths from snakebite in the U.S., far fewer than are caused by insect stings.

This book presents reliable scientific information about the common reptiles and amphibians found in North America north of Mexico. It will give you the facts to understand and appreciate these intriguing and diverse animals.

REPTILES include three groups: turtles, lizards and snakes, and alligators and crocodiles. (The tuataras, a fourth group, are found only in New Zealand.) Reptiles are believed to have evolved from amphibians around

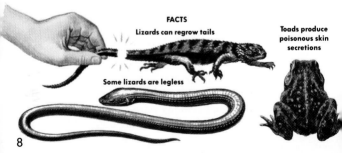

FACTS

Lizards can regrow tails

Toads produce poisonous skin secretions

Some lizards are legless

8

250 million years ago (see pp. 4–5). They are found on every continent, except Antarctica, living in habitats ranging from swamps to deserts. This diversity is one of the reasons they have managed to survive for so long.

Reptiles in this book range in size from ground lizards (about 2 to 3 inches long) to alligators (12 feet or more). Most have scales that form a waterproof barrier around their bodies. This protects them from dehydration, whether they live on dry land or in water. Most reptiles lay their eggs on dry land, but some incubate their eggs within the mother's body and bear their young alive.

AMPHIBIANS include frogs, toads, and salamanders. Their ancestors appeared about 300 million years ago (see pp. 6–7). Most spend at least part of their lives in water. Most live in moist areas. The populations of many species are declining today (see below), often because of draining and developing wetlands for human use.

Amphibians in North America range from tiny tree-frogs (less than 1 inch long) to Amphiumas (up to 3 feet). Most have smooth skin, some with a moist coating. Amphibians lay their eggs in water, often in jelly-like masses.

ENVIRONMENTAL VALUE Reptiles and amphibians are an important part of the environment of many habitats. They help control harmful pests and are prey for

FABLES

Snakes roll like hoops

Snakes hypnotize birds

Toads cause warts

9

other animals. Their success or failure may also be a measure of the health of an ecosystem and reveal how human activities are affecting the environment.

The populations of many species of both reptiles and amphibians (except possibly the Bullfrog, p. 133) are declining significantly. Some have become extinct. Biologists suspect that the problem may be global. In addition, the offspring of many species of amphibians have unusual malformations such as deformed limbs. These changes may be related to human activities, including land development, pollution, and the use of pesticides. If you find a malformed amphibian, notify the North American Reporting Center for Amphibian Malformations (NARCAM) at 1-800-238-9801. Their website is found at http://www.npsc.nbs.gov/narcam/.

CONSERVATION Needless killing, often based on fear and misunderstanding, must stop. Venomous reptiles found near human habitations should be removed by competent authorities. Wild areas where reptiles, amphibians, and other wildlife live should be preserved. There are a number of programs for interested volunteers. For information contact the North American Amphibian Monitoring Program (NAAMP), which maintains a website at: http://www.mp1-pwrc.usgs.gov/amphibs.html. They manage counts of eggs, tadpoles, and adults and calling surveys that record the number and kinds of peeps or croaks heard during breeding season.

Partners in Amphibian and Reptile Conservation (PARC) is the largest organization focused on conservation of both reptiles and amphibians. This nonprofit organization includes federal and state agencies, conservation groups, universities, zoos, private industry, and individual members from the general public. Anyone can join PARC at http://www.parcplace.org.

RECOGNIZING REPTILES AND AMPHIBIANS
Learning to know reptiles and amphibians from books like this one can save you time and effort in the field. Study those found in your region first. Zoos and museums can help. Learn to recognize venomous snakes at a glance.

MAPS In this book the approximate ranges of North American species found north of Mexico are shown on maps. They can help you identify which species are most likely to be seen in your area. Where a map shows ranges of more than one species, the common name of each appears within or next to the color or hatching that shows its range. Overlapping colors and crosshatching indicate overlapping ranges.

SCIENTIFIC NAMES Although common names are used throughout this book, the scientific name for each species illustrated is given on pages 155–157. The scientific name consists of two words—first the generic name (genus) and then the specific name—the two together denote a species. A third name indicates a subspecies.

**Hunting Frogs
with Headlight**

Snake Collecting Bag

Snake Stick and Noose

OBSERVING IN THE FIELD

Familiarize yourself with places where reptiles and amphibians are found.

The life histories of many reptiles and amphibians are still unknown. Eating habits, wintering habits, and mating habits of many species are mysteries. Sometimes the adults have been described, but we know little about their eggs or young. First study the animals carefully in the field. The more natural the conditions, the better your observations will be. Binoculars are often a help and a notebook is essential. You may also want to observe a reptile or amphibian in captivity for further details. If you do collect harmless species, turn them loose after you have studied them.

COLLECTING EQUIPMENT

A strong net will help you capture amphibians, though some collectors prefer to grab them by hand. A snake

Terrarium for Frogs

stick can be used to pin down a snake. Leave venomous snakes alone. Carry reptiles in cloth bags or pillow-cases. Even when the open end is tied, they usually pro-vide enough ventilation, but a bag can be a death trap for a specimen if it is left in the hot sun or in a closed-up car. Cans or plastic jars with air holes are fine for amphibians. Keep moist leaves or grass in the container to prevent their skin from drying out.

CAGES AND TANKS Keep amphibians in aquariums. Some require a rock or float so they can climb out of the water. Others, especially tadpoles, can use any aquarium suitable for fish. Toads, however, need a moist terrarium. Lizards and snakes can be held in a wooden cage made of plywood with a glass front. Allow at least a square foot of floor space in the cage per foot for a medium-sized snake, more for larger species. The top of the cage, the door, should be hinged. Three or four one-inch (or larger) holes at the ends and back of the

Amphibian Eggs in Aquarium

cage will aid ventilation. Cover these holes tightly with fine screen.

Try to duplicate the natural habitat of the lizard or snake. Put sand, gravel, or newspaper on the floor of the cage. Add a rock or two and a dish of clean water that is large enough for your snake or toad to soak in. Attach the water container so a moving animal cannot turn it over. For lizards and snakes that climb, provide a larger cage and set a branch in it. Be sure that the floor of the cage is always dry. Reptiles kept in wet cages often develop skin infections which are difficult to cure. Sick reptiles or amphibians should not be released as some states have laws prohibiting the release of animals, but should be euthanized and disposed of, or given to a museum for preservation.

Holding Snake Safely

KEEPING REPTILES AND AMPHIBIANS It is relatively easy to keep reptiles and amphibians in captivity, but many species cannot be possessed legally without a permit. Inquire at your state Wildlife Department and consult herpetologists before you attempt to keep any reptile or amphibian caught in the wild, or before you kill a venomous species. Captured animals should be returned to where they were found. Venomous species should never be kept.

WHAT IF IT BITES? Get to medical treatment as quickly as possible. While on the way to a doctor, most experts advise washing the bite with soap and water and immobilizing the bitten area; keep it lower than the heart.

If unable to reach medical care within 30 minutes, wrap a bandage two to four inches above the bite to slow the venom. The bandage should be loose enough so a finger can be slipped under it.

WHAT NOT TO DO

• Don't wait to get medical help.
• Don't try to cool the bite with ice or any other type of cooling material.
• Don't use a tourniquet.
• Don't cut the wound.

Certain venomous snakebites may be treated with antivenin. This should only be done by a doctor. The patient needs to be monitored at a medical facility.

Attempts at first aid often do more harm than good. If given at all, it should be at once.

Apply constriction band, not a tourniquet.

Keep patient quiet, warm, and comfortable.

Phone or transport to doctor immediately.

TURTLES
(49 U.S. Species)

LIZARDS (95 spec

Scaly Skin

Plated Skin

REPTILES (teeth alike)

REPTILES (the class reptilia) found in North America are divided into three orders. *Chelonia* includes turtles and tortoises (pp. 18–43) with their hard shells. *Squamata* contains lizards (pp. 44–69) and snakes (pp. 70–113). Lizards are generally small and varied. Snakes are closely related, but have no limbs. *Crocodilia* includes alligators and crocodiles (pp. 114–115), which are generally the largest of the reptiles.

All reptiles, even aquatic species, have lungs and breathe air. Their skin is usually covered with scales or plates made out of a substance called keratin. Those species that have toes usually have claws. Most reptiles lay eggs. In a few species, the eggs develop inside the mother and the young are born alive. All young are miniature versions of their parents, although they may have different coloration. They are able to take care of themselves soon after birth or hatching.

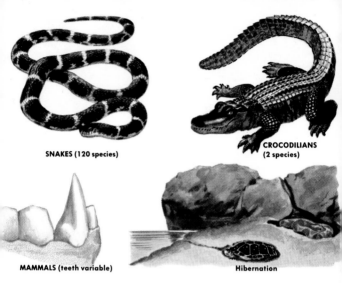

SNAKES (120 species)

CROCODILIANS (2 species)

MAMMALS (teeth variable)

Hibernation

Reptiles are said to be cold-blooded (ectothermic). This means their body temperature is the same as the temperature of their surroundings. Warm-blooded (endothermic) creatures use calories from food to regulate their body temperature. Instead, reptiles control their temperature through behavior. Many bask in the sun to warm up. Some cool themselves in water or avoid direct midday sun. During the hottest weather some reptiles become dormant (aestivate). In cooler regions they may also become dormant to avoid the cold (hibernate). In North America many hibernate from late fall to early spring under soil, rocks, or water.

Reptiles are not always easy to find. Some are small, many are nocturnal, and most have protective coloring. Almost all feed on rodents and insects and play an important role in controlling pests. Only a few snakes and lizards are venomous; the great majority are harmless.

BOX TURTLE

TURTLES were long thought to be the oldest surviving group of reptiles. Their ancestors first appeared some 200 million years ago, during the time of the dinosaurs. Many of the living families of turtles have undergone little apparent structural change since then.

Turtles are easily recognized by their hard, armorlike shell. The top shell, or carapace, covers the turtle's back and sides. The lower shell, or plastron, protects its belly. On all turtles, the two parts are attached at the sides. A turtle's legs are connected within their shell. This allows them to retract their legs for protection. When startled, most turtles also withdraw their heads straight back into their shells, folding their neck into a tight S-shaped bend.

Turtles have no teeth, but their horny bill can slice through plant or animal food. A few species are largely herbivorous, but most eat insects, worms, grubs, shellfish, and fish. All turtles lay eggs and bury them on land. Most lay 6 to 12, but Sea Turtles lay many more.

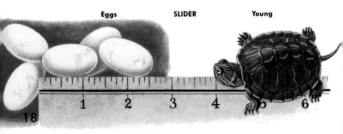

Eggs   **SLIDER**   Young

In many species, sex is determined by the temperature at which the eggs develop. After hatching it takes 5 or more years for young turtles to grow to maturity. Male turtles are generally smaller than females, but they have a longer tail and many have a concave plastron.

Turtles are found throughout the temperate and tropical world and in the open ocean. In colder areas, turtles hibernate through winter under soil or in the mud at the bottom of ponds. Some also become dormant in hot, dry weather.

Of the 270 or more known species of turtles, more than 100 are considered rare or threatened with extinction. Living turtles of North America and adjacent seas fit into seven families. Six are illustrated at the right by representative species. A member of the seventh family, land tortoises, is pictured on page 27. This is the only group correctly called "tortoises."

LEATHERBACK

GREEN TURTLE

MUD TURTLE

SNAPPING TURTLE

SOFTSHELL TURTLE

FALSE MAP TURTLE

LEATHERBACK

HAWKSBILL

SEA TURTLES are larger than, and different from, pond and land species. The limbs of marine turtles are modified into flippers—streamlined for swimming, clumsy for use on land. As a result, these turtles seldom come ashore, though the female does so to lay her large batch of eggs in late spring. The eggs are buried in the sand just past the high-water mark. Sea Turtles are found in warmer waters of both Atlantic and Pacific, and occasionally off northern shores in summer. Of five kinds, the Leatherback is largest. Specimens over 8 ft. long, close to 1,500 lb., have been

LOGGERHEAD

GREEN TURTLE

caught. The ridged, leathery back makes identification easy. The Hawksbill, smallest of the Sea Turtles, also is easy to recognize because of its overlapping scales. This is the species from which "tortoise shell" comes. The Green Turtle has four plates on each side between the top and the marginal plates. The Loggerhead and Ridleys (not illustrated; two species) have five to seven plates on each side. The Loggerhead is much larger than the Ridleys and usually has three scales at the sides of the plastron; the Ridleys have four.

MUSK TURTLES are aquatic species of ponds, slow streams, and rivers. They often sun themselves in shallow water, but seldom come ashore. The females do so to lay eggs. Note the narrow, high carapace, often covered with algae and water moss. The lower shell (plastron) is narrow and short, almost like that of Snapping Turtles. The Musk Turtle has a strong odor. Four species occur; the commonest, shown above, has two light stripes on each side of its head.

22

EASTERN MUD TURTLE

YELLOW MUD TURTLE

MUD TURTLES, five species of them, live about the same as Musk Turtles. They are aquatic, feeding on larvae of water insects and small water animals. Notice that the plastron is much wider in the Mud Turtle and is all scaly. Both ends are hinged, so that the Mud Turtle can pull the plastron in, giving head and limbs more protection. Mud Turtles have a musky odor, too. They are small, rarely over 4 in. long, and are more common in the Southeast.

23

SNAPPING TURTLE and its giant relative (p. 25) are dangerous. Their long necks, powerful jaws, and vicious tempers make them unsafe to handle. Experts carry them by the tail, well away from the body. Snappers are aquatic, preferring quiet, muddy water. They eat fish and sometimes waterfowl. Note sharply toothed rear edge of the rough carapace, often coated with green algae. Plastron is small. Adults, to 18 in. or more, weigh 20 to 62 lb (86 in captivity).

ALLIGATOR SNAPPING TURTLE is the largest freshwater turtle, reaching a length of 30 in. and a weight of close to 235 lbs. Entirely aquatic, it lies on the muddy bottom, its huge mouth agape, wiggling a pink, wormlike growth on its tongue to attract unwary fish. Its powerful jaws can maim a hand or foot. It differs from Common Snapper in having three high ridges or keels on its back. Specimens are reported to have lived 75 years and more in zoos.

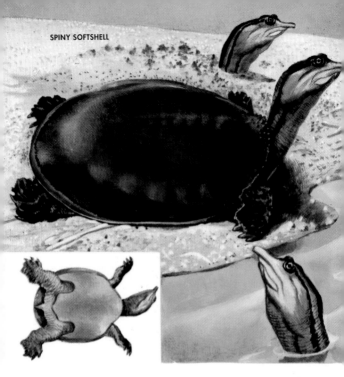

SPINY SOFTSHELL

SOFTSHELLS have, in fact, a hard shell, but it is soft-edged and lacks horny scales. These turtles pull in their head and limbs for protection nevertheless. Of three species, two have small bumps or tubercles along the front edge of the carapace; the other does not. All have long necks, sharp beaks, vicious tempers. Handle them by rear of shell. These turtles grow to a length of about 20 in. and weigh up to 35 lbs. All species are aquatic, although they often bask on the shore.

GOPHER TORTOISE

TORTOISES are land turtles with blunt, club-shaped feet very different from the webbed feet of aquatic species. Their diet includes much plant material as well as insects and small animals. Our three species differ externally in minor ways but are placed in two genera. Their family includes the Giant Tortoises of the Galapagos Islands, largest of land turtles. The high, arched carapace and the habit of burrowing are characteristic except in the Texas species.

27

ELEGANT SLIDER
(melanistic or dark male)

Front
Claws
of Male

Claws
of Female

SLIDERS are a common group of six species. The carapace is usually smooth and fairly flat, the rear edge roughly toothed. The carapace of the Florida and Alabama species arches higher than the carapace of others. The olive-brown shells and skins of Sliders are splotched with red and yellow. The Red-eared Slider has a distinctive dash of red behind the eye. The males, much darker than females, were once mistaken for different species. During courtship they seem to tickle or gently scratch the female's head with the extra-long toe-

RED-EARED SLIDER

Female and Young

nails on their front feet. The female later digs a hole and deposits about 10 eggs, which she covers with dirt.

On warm days Sliders sun themselves on logs or debris. They often stack themselves two or three turtles high, but the whole pile will plunge into the water if frightened. Sliders eat mostly vegetation. They live more than 30 years, growing to about 1 ft., and are the most common turtles in the South.

HIEROGLYPHIC RIVER COOTER

COOTERS have a dark, flattened carapace, marked with yellow. Their plastron is yellow with dark markings. One species has been named the Hieroglyphic River Cooter because the markings on its shell and skin resemble ancient writing. Cooters grow to 10 to 12 in. long; females are larger than males. Cooters feed on water plants, small water animals, insects, and even dead fish. They often bask.

CHICKEN TURTLE is small (5 to 8 in.) with an exceptionally long neck and a pattern of narrow yellow lines on its brownish carapace. Its shell is higher and narrower than a slider's, and it has a smooth rear edge. A Chicken Turtle has yellow stripes on its legs, head, and neck, and its plastron is yellow. It prefers the still waters of ditches and ponds and often hibernates on land.

EASTERN PAINTED TURTLE

PAINTED TURTLES are perhaps the most common and widespread of turtles. They are found wherever there are ponds, swamps, ditches, or slow streams. These small (5 to 6 in.) turtles spend much of their time in or near water, feeding on water plants, insects, and other small animals. They are also scavengers. In summer, Painted Turtles gather together, and if one approaches quietly, they may be seen sunning on logs, rocks, or even floating water plants. Males are similar

SOUTHERN PAINTED TURTLE   WESTERN PAINTED TURTLE

MIDLAND PAINTED TURTLE

to the females but smaller, with the same long nails on their forefeet that Sliders have. Females lay 6 to 12 white eggs in a hole dug laboriously with their hind legs in the soil. The eggs may hatch in two or three months, though some young do not emerge till the following spring. Painted Turtles may be easily identified by their broad, dark, flattened, smooth-edged shells. The margin of the carapace is marked with red; so is the yellow-streaked skin, especially on the head and limbs. The plastron is yellow, sometimes tinted with red. In all four subspecies of Painted Turtles the upper jaw is notched in front. The notch has a small projection on each side. Markings and details of carapace and plastron differ from subspecies to subspecies. Painted Turtles are shy and are not easily captured. Hardy and adaptable, they survive well even in urban areas and in harsh cold. Though frantic when first captured, the turtles are not aggressive.

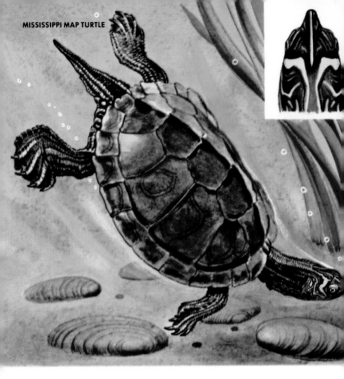

MISSISSIPPI MAP TURTLE

MAP TURTLES are aquatic turtles often found in large numbers in ponds, swamps, and quiet streams. They are even more timid than Painted Turtles. Differences between the sexes are more extreme than in other turtles, females reaching nearly 13 in., males less than half that. Females develop a grotesquely broad head with massive crushing jaws, adapted for feeding on hard-shelled clams and snails. Males and juveniles feed on soft-bodied insects and other aquatic animals. Males seek shallow, debris-laden waters, often sunning them-

selves; females remain mostly in open, deep, muddy-bottomed waters and seldom sun themselves. The female, coming ashore briefly in early summer to lay 10 to 16 eggs, returns to the water as soon as the eggs are buried. Map Turtles (ten species) are named for the faint yellow pattern on the carapace. Lines are brighter on head and limbs. The keeled carapace and its roughly toothed rear edge are identification marks.

Map
Others
Mississippi

35

BLANDING'S TURTLE with its hinged plastron somewhat resembles the Box Turtle, but cannot close its shell tightly. It has webbed feet and lacks the hooked bill of the Box Turtle. The plastron is notched at the back. Reaches a length of 10¼ in., but is commonly 7 to 8 in. Prefers quiet waters, but also lives in marshes, where it feeds almost entirely on crayfish and insects. Yellow and black markings make this shy species especially attractive. It is becoming rare.

DIAMONDBACK TERRAPIN gets its name from the intricate pattern of markings and rings on its carapace. The plastron can be yellow, mottled, or dark, and its head and legs are usually spotted. Terrapins are found in the brackish water of coastal marshes and tidewater streams. They feed on small shellfish, crabs, worms, and plants. Females are larger than males, growing from 6 to 9 in. long.

37

EASTERN BOX TURTLE

BOX TURTLES are land species, occasionally found in or near water, though they are well adapted for life on land. They prefer moist, open woods or swamps and feed on insects, earthworms, snails, fruits, and berries. Box Turtles have a hinged plastron which they pull tight against the carapace for complete protection when they are frightened. The carapace, 4 to 5 in. long, is highly arched. Of the two species of Box Turtles, Eastern and Ornate, the former is divided into four subspecies, the latter into two, distinguished by the shape and markings on the shells and by the number

EASTERN BOX TURTLE

ORNATE BOX TURTLE

ORNATE BOX TURTLE

of toes (three or four) on the hind feet. The plastron of the female is usually flat; that of the male, curved inward. Males have longer tails, and their eyes are usually bright red. The female has dark reddish or brown eyes. In early summer the female buries four or five round, white eggs in a sunny spot. These hatch in about three months. The young may hibernate soon after, without feeding. Young Box Turtles grow ½ to ¾ in. yearly for five or six years; then they grow more slowly—about ¼ in. a year. At 5 years they mate and lay eggs; at 20 they are full-grown, and they may live to be as old as 40. Box Turtles have been reported living 25 years and more in captivity. Though seemingly docile, some bite unpredictably. They eat poisonous mushrooms without harm, but the poison is stored in their flesh, which when eaten by other animals (including humans) can cause death.

Ornate

Eastern

39

SPOTTED TURTLE is a small (3 to 5 in.) common spring turtle with round orange or yellow spots on its smooth, black carapace. The head is colored similarly. The young have but one spot on each scale, or none. Living in quiet fresh water, this turtle feeds on aquatic insects, tadpoles, and dead fish, but eats only when in water. The tail of the male is about twice as long as the female's. Usually three eggs are laid in June.

40

WESTERN POND TURTLE is related and similar to the eastern Spotted Turtle, but larger—6 to 7 in. The yellow dots and streaks on the carapace are faint. The plastron, concave on the male, is yellow with dark patches at the edges. This is the only freshwater turtle of the far West. Living in mountain lakes, marshes, and in slow stretches of streams with abundant aquatic vegetation, Western Pond Turtles feed on small water life, including some plants.

BOG TURTLE, smallest in the world, is quickly identified by large orange spot on each side of head. Dark carapace is short (3 to 4 in.) and narrow, marked with concentric rings. This turtle is semiaquatic, living in mud-bottomed bogs, swamps, and slow streams, feeding omnivorously. In June–July, three to five eggs are laid. As its habitat has been drained and developed, the Bog Turtle has become rare or absent from much of its range. It is federally protected.

WOOD TURTLE is easily recognized by its deeply grooved, rough shell, which has earned it the nickname "sculptured turtle." It is also called "redleg" because of its orange-red skin. Wood turtles spend much time on dry land, especially in moist woods. They move to open land to feed and to swamps, ponds, and slow streams when the weather is dry. They are omnivorous, nest in May to June, and lay 4 to 12 eggs. Adults are 7 to 9 in. long.

43

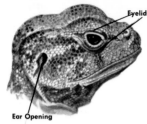

Eyelids

Ear Opening

LIZARDS are generally small and very diverse. They are found mainly in the warmer parts of the world, although a few species live as far north as Canada and Finland. Lizard fossils have been found in rocks formed during the period when dinosaurs were common. Today over 5,000 species are known, outnumbering all other reptiles. They are grouped into about 26 families, 11 of which occur in the United States. More than 400 species are found in North America, of which more than 120 live within the boundaries of the United States, most in the South or West.

Lizards are closely related to snakes, but there are significant differences. Lizards are typically four-legged, with five toes on each foot. They have scaly skins like snakes, but they molt in flakes rather than shedding their whole skin at once. Lizards usually have movable eyelids and ear openings on the sides of the head. Snakes have neither. Even those few species of lizards that have long bodies and no limbs (pp. 67–68) can be easily distinguished from snakes. They have several or many rows of scales on their underside. Snakes usually have only a single row. Salamanders (pp. 137–153) are also sometimes mistaken for lizards, but they live in moist places, have smooth skins, no claws,

Belly Scales—LIZARDS

Belly Scales—SNAKES

and fewer than five toes on their front feet.

Most lizards lay eggs, burying them in the ground, but the eggs of a few species develop inside the mother's body and the young are born alive. Males and females look alike in some species, but in many they differ in size and color. A few species feed primarily on plant material, but most lizards feed on insects and other small animals. They recognize their prey by its movement and grasp it with lightning speed. Most are valuable in controlling insects.

Lizards are not easy to catch. They can run rapidly—the fastest has been clocked at about 15 miles per hour. Most lizards are active during the day, often basking in the sun. Many can swim and some desert species use a swimming motion to move just below the surface of sand. Males of most species are intolerant of each other, especially during the breeding season. Females are less aggressive. In the wild, some kinds may live over 50 years; others average less than two years.

CENTIPEDE

FLY

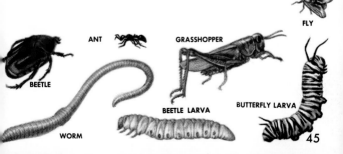

BEETLE

ANT

GRASSHOPPER

BEETLE LARVA

BUTTERFLY LARVA

WORM

45

LEAF-TOED GECKO

GECKOS are unusually attractive lizards, recognized by their large, often lidless eyes with vertical pupils. The skin, usually covered with fine, beaded scales, is almost transparent. Most have enlarged, padded toes. Some live around houses, others on rocks or trees, feeding on small insects. They are nocturnal. Geckos lay one to three or more small white eggs with brittle shells during summer. Five or more tropical species have become naturalized in Florida. Most are docile and rarely bite. The tail breaks off easily. Leaf-toed and Banded Geckos are our only truly native species. The Dwarf Gecko long ago came to Florida from the West Indies, the Turkish Gecko recently from North Africa.

ASHY GECKO

MEDITERRANEAN GECKO

WESTERN BANDED GECKO

BANDED GECKOS (four similar species) are western lizards (4 to 5 in. long). Tails are about half the body length but have usually broken and regrown shorter. Banded Geckos live in rocky or sandy deserts and lower mountain slopes. They come out at night to feed on small insects and, in turn, are eaten by snakes and larger lizards. They hibernate from Oct. to May but are common other months. They never bite, but are shy, quick, and easily injured.

Banded

Leaf-toed

Other Geckos

47

GREEN ANOLE (male)

GREEN ANOLE (female)

TRUE CHAMELEON

GREEN ANOLE, common and attractive, can change color from dark brown to bright green. It is sometimes mistakenly called a chameleon, but true chameleons are Old World lizards. Male anoles have a pink flap of skin on their throat shaped like a fan. Eggs are laid singly every two weeks in summer under moist debris; they hatch in about six weeks. At least six other species of anoles have become established in Florida from the West Indies.

CHUCKWALLA (16 in. long) is our second largest lizard, after the Gila Monster (p. 69). It feeds mainly on the tender parts of desert plants, including cactus. Chuckwallas hide in crevices when disturbed. They sometimes make a sound like rubbing sandpaper as they slide under. If bothered further, they gulp air and inflate their bodies, wedging themselves in the rocks. Clutch of 3 to 8 eggs laid June to August.

DESERT IGUANA, a handsome spotted species of sandy deserts, lives in burrows under sparse shrubs. It feeds mostly on tender desert plants, sometimes climbing up them to reach its food. The Desert Iguana is fairly large (12 to 15 in.), with a tail almost twice as long as its body. It is wary and runs rapidly, often taking refuge in its burrow. Each has its own territory for feeding. Female lays 3 to 8 eggs in July or August.

50

TRUE IGUANA

SPINYTAIL IGUANA

SPINY and TRUE IGUANAS, representing two groups of large American lizards, are not found natively in the U.S., but come to within 100 miles of our border and have established colonies through introduction in Florida, Texas, and California. About 10 or 11 species of Spiny or False Iguana (1 to 4 ft. long) live in Mexico and Central America. The True Iguana (4 to 6 ft.) lives in the same area, mostly in trees. Both are locally favorite foods. Both are herbivorous. Other Iguanas live in the Galapagos Islands and West Indies.

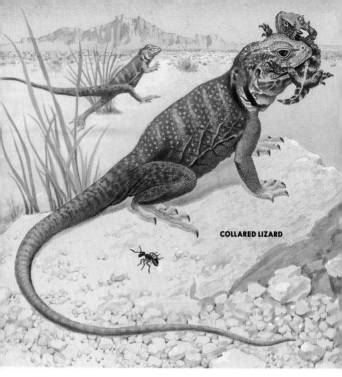

COLLARED LIZARD

COLLARED LIZARDS The black collar marks the Collared Lizard (four species); so does its long tail, plump body, thin neck, and relatively large head. Males are more brightly colored, with a tinge of orange and yellow. Collared Lizards, fairly common in rocky areas, feed on insects and small lizards. Wary, they can run swiftly on their hind legs. Collared Lizards bite when captured, lay 1 to 2 clutches of 1 to 13 eggs. A species of the lower Rio Grande valley lacks the black collar. A cornered animal gapes the mouth widely, showing its jet black lining.

COLLARED
LEOPARD LIZARD

LEOPARD LIZARDS resemble Collared Lizards, but are more spotted and have narrower heads and bodies. They prefer flat sand or gravel areas with sparse vegetation. They eat insects and lizards, often their own kind. Like Collared Lizards, they use only their hind legs when running fast. Females develop a deep salmon color on their undersides during breeding season. They lay 1 to 2 clutches of 1 to 11 eggs, which hatch in a month or so. They will bite if handled.

Leopard
Collared

53

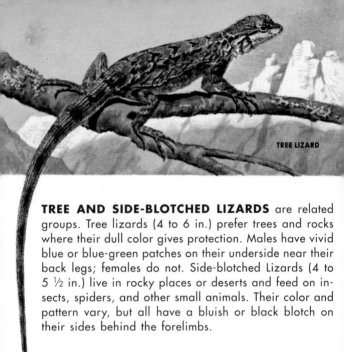

TREE LIZARD

TREE AND SIDE-BLOTCHED LIZARDS are related groups. Tree lizards (4 to 6 in.) prefer trees and rocks where their dull color gives protection. Males have vivid blue or blue-green patches on their underside near their back legs; females do not. Side-blotched Lizards (4 to 5 ½ in.) live in rocky places or deserts and feed on insects, spiders, and other small animals. Their color and pattern vary, but all have a bluish or black blotch on their sides behind the forelimbs.

Side-blotched

Tree

SIDE-BLOTCHED
LIZARD

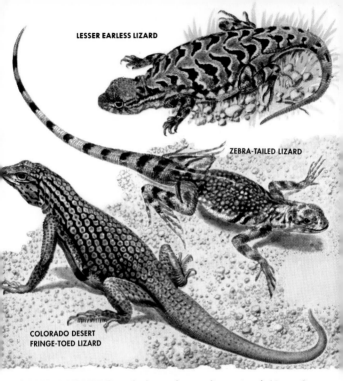

LESSER EARLESS LIZARD

ZEBRA-TAILED LIZARD

COLORADO DESERT
FRINGE-TOED LIZARD

SAND LIZARDS include eight medium-sized (6 to 8 in.) lizards (of four genera), all preferring sandy terrain. Two of the species above are at their best in the sand dunes of the California and Arizona deserts. All have a skin fold across the underside, in front of the forelegs. Legs and fringed toes are long. Tails, about body length, are often marked with black bars underneath. Lip scales slant, help in sand-burrowing. All feed on small insects.

Zebratail
Lesser Earless
Fringe-toed

TEXAS SPINY LIZARD

WESTERN FENCE LIZARD

SAGEBRUSH LIZARD

CREVICE SPINY LIZARD

DESERT SPINY LIZARD

SPINY LIZARDS form a large group of common lizards, including Fence, Spiny, and Sagebrush Lizards. Some 37 forms (species and subspecies) live in the U.S., almost three times as many farther south. The largest have bodies about 5 in. long, tails slightly longer. All are active in daylight, spending the night in cracks, crevices, or on branches. Some species lay eggs; others bear 6 to 12 young alive. Head, body, and limb forms are guides to the entire group, once you learn them. These lizards lack the skin fold across the throat

Underview

Clutch of Eggs

EASTERN FENCE LIZARD

that Sand Lizards (p. 55) and similar species have. Some Spiny Lizards are blue or blue-patched on the underside; this is more pronounced in males. Detailed identification may be difficult. Spiny Lizards are good climbers; they are often found in trees, on boulders, among rocks. Their food is mainly small insects. Two species live in Florida, one elsewhere in eastern U.S. Others occur from Texas and Wyoming westward.

Spiny Lizards
(16 species)

DESERT HORNED LIZARD

SHORT-HORNED LIZARD

HORNED LIZARDS are unique. These odd, flattened creatures are found only in the West and in Mexico. The only other lizard like them is one in Australia. All have various-sized spines on the head which give the group its name. Seven species are found in dry, sandy areas, where they lie on rocks or half buried in the sand. When an insect appears, a quick snap of the lizard's tongue takes care of it. Some species lay 20 to 30 eggs; in others up to a dozen young are born alive. In one species eggs hatch in only a few hours; others take several weeks. These unusual liz-

TEXAS HORNED LIZARD

Underground Clutch of Eggs

ards may squirt a thin stream of blood from the corners
of their eyes when frightened. Some puff up when angered; others flatten themselves out even more. Horned
lizards hiss threateningly, jump at an intruder, lift themselves high on their legs, and tilt their body toward
danger. All such behavior is a bluff,
however; the animals rarely bite.
Protected in most states, they should
never be kept as pets. Ants are their
favorite food, but they do not eat
well in captivity.

59

GRANITE NIGHT LIZARD

DESERT NIGHT LIZARD

NIGHT LIZARDS are mottled, medium-sized lizards. Both body and tail are about 3 in. long. They live in areas of granite, behind the loose-scaled flakes of rock or under fallen stalks of yucca. Note the vertical pupil in the eye and lack of eyelids. Horizontal rows of

plates cross the belly. The three species are nocturnal, spending the day in sheltered cracks. Young (two or three at a time) are born alive. The food is beetles and other small insects.

SKINKS are found in most parts of the United States; no other lizards have as wide a range. All are small to moderate in size. Few have bodies more than 5 in. long or tails much over 6 in. Skinks can be recognized by their smooth, flat scales, which give them a glossy, silky appearance. Most have short legs but are swift runners. Most are ground lizards, although a few are arboreal. All burrow occasionally. Active during warm days, they feed on insects, spiders, and worms. Skinks are usually found in damp habitats, but they can also be found in vacant lots, woods, or meadows. Skinks hibernate through winter and most mate in spring. From 6 to 18 eggs are laid about six weeks later. The mother spends the next six or seven weeks brooding her eggs until they hatch, an unusual behavior for lizards. Newly born skinks are about an inch long.

Skinks can be roughly identified by the markings on their backs. Most common in the East are five-lined skinks, which have five light lines running down the body; many have bright blue tails. In the West, four-lined skinks are common, but other skinks have eight lines, two lines, or no lines at all. Larger species have powerful, narrow jaws and can inflict painful bites. Their body often thrashes vigorously while the jaws maintain their grip. A skink's tail will break off very easily if it is grasped. It will continue to wriggle for some time, often distracting a predator.

Skinks
(15 species)

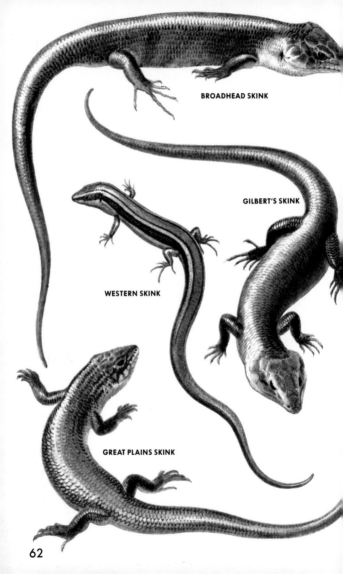

BROADHEAD SKINK

GILBERT'S SKINK

WESTERN SKINK

GREAT PLAINS SKINK

62

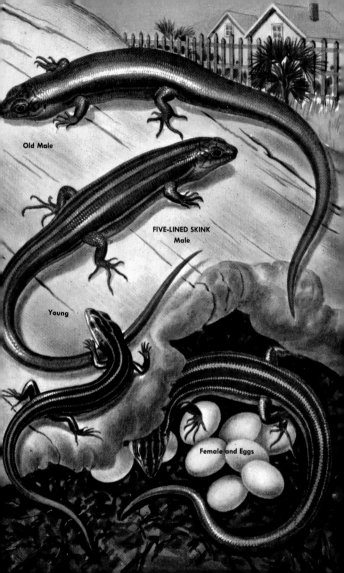

Old Male

FIVE-LINED SKINK
Male

Young

Female and Eggs

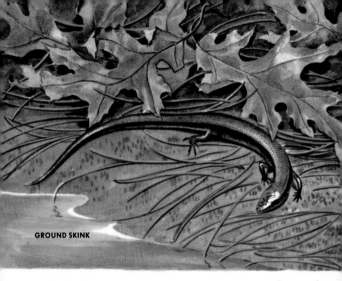

GROUND SKINK

GROUND AND SAND SKINKS are similar to the skinks on the previous pages. The Ground Skink (2 to 4 in. long) is golden brown to dark brown with darker brown stripes down its sides. It has an unusual transparent disk in its eyelids that allows it to see with its eyes closed. Usually found on the ground, it hunts insects in leaves or rotted wood. If disturbed, it hides under the nearest shelter, running with a sidewise, snakelike movement. The Sand Skink (4 in. long) is a burrowing lizard found in dry, sandy soils in pine woods. It "swims" through sand, seldom using its small legs. It feeds on termites and beetle grubs.

SAND SKINK

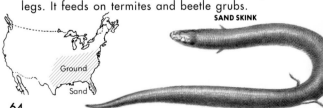

Ground

Sand

SIX-LINED RACERUNNER

WESTERN WHIPTAIL

WHIPTAILS AND RACERUNNERS look so much alike that even experts have trouble telling species apart. The common and widespread Racerunner is about 3 in. long with a tail twice that length. These active lizards are found in dry, open areas, feeding during the day on insects, worms, and snails. Whiptails are checkered or striped. Most are found west of the Great Plains.

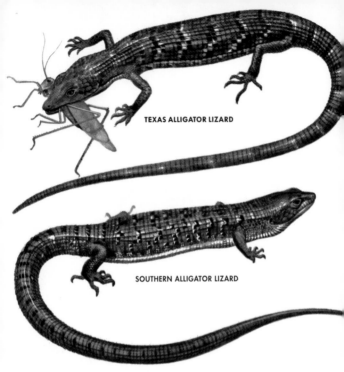

TEXAS ALLIGATOR LIZARD

SOUTHERN ALLIGATOR LIZARD

ALLIGATOR LIZARDS, named for their shape and heavy scales, are slow, dull-colored, solitary, with a banded or speckled back. Fairly large (to 8 in., tail twice as long if complete), they feed on insects and spiders and, in turn, are food for larger reptiles, mammals, and birds. A skin fold with tiny scales along sides of body allows for expansion of the abdomen with food or eggs. Males may bite painfully. Five species: four lay eggs; in one, young are born alive.

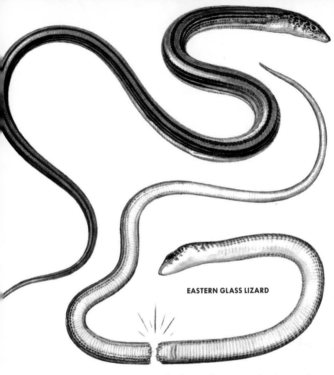

EASTERN GLASS LIZARD

GLASS LIZARDS (2 to 3 ft. long) have no limbs and are somewhat snakelike. Ear openings, eyelids, and many rows of belly scales proclaim them to be true lizards. The very long tail breaks off more easily than that of other lizards. It may break off when the animal is captured or roughly handled. The tail, of course, cannot rejoin the body, but a new, shorter tail grows in its place. These lizards feed on insects. They may bite when handled.

FLORIDA WORM LIZARD

CALIFORNIA LEGLESS LIZARD

WORM and LEGLESS LIZARDS are two small burrowing species. The former (up to 10 in. long, only ¼ in. thick), found in sandy soil of pine woods, has distinct rings which make it look much like a large earthworm. Limbless, earless, and blind, it belongs to a tropical

Legless

Worm

suborder of a rank equal to those of lizards and snakes. The California Legless Lizard, which is even smaller (6 in.), has small eyes but is earless and limbless. One form is silvery, the other black. All eat small insects.

GILA MONSTER, our only poisonous lizard, grows to 2 ft. long. The poison comes from glands in the lower jaw. Usually slow and clumsy, Gila Monsters can twist their heads, bite swiftly, and hang on strongly, chewing the poison into the wound. It is illegal to capture or molest them. Gila Monsters live under rocks and in burrows by day. They feed on eggs, mice, and other lizards. The 6 to 12 eggs hatch in about a month.

69

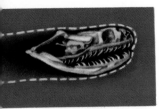

Skull of Non-poisonous Snake—
Eastern Racer

Skull of Poisonous Snake—
Cottonmouth

SNAKES are closely related to lizards. More than 100 species are found in the U.S.; 19 produce venom that can seriously harm humans, but death from snakebite is an extreme rarity. Most snakes live on the ground, but some live underground, in trees, or in water. They are generally solitary, although they may congregate where there is food or shelter, or to hibernate or mate.

Snakes have elongated bodies and lack limbs. Their skin is covered with horny scales and is usually protectively colored. Except for Blind Snakes, all have a single row of large scales (small in U.S. boas) down their belly. Snakes shed their skin once or several times a year.

Snakes have no vocal chords, but can hiss. They also lack ear openings and movable eyelids. Their eyesight is fairly good, though their distance vision is poor. They are very sensitive to smells. A snake's long, forked tongue con-

Copperhead Shedding

Head of Scarlet King
Snake (showing tongue
attachment and teeth)

Eggs of Red King Snake

Copperhead at Birth

stantly tastes its surroundings, bringing particles into contact with a smell-sensitive organ in its mouth. Snakes do not react to airborne sound waves as we do; they respond to vibrations through the ground.

Snakes feed on live animals, including insects, worms, frogs, rats, mice, and birds. Each side of a snake's lower jaw can move separately, enabling it to swallow prey larger than its normal mouth size. Their teeth are small and hooked. The larger fangs of venomous species are grooved or hollow.

Some snakes lay eggs; others bear their young alive. Many have about a dozen young at a time, but some have as many as 100. The young look like miniature versions of their parents and can fend for themselves. Most double their size in the first year and are full grown in two or three years.

Snakes are important controllers of pests, especially rats and mice. Snakes can be kept in captivity if they are properly cared for.

Shed Skin

WESTERN BLIND SNAKE

BLIND SNAKES, wormlike in color and size (8 to 12 in.), are truly blind. They may come to the surface at night. Most are found under stones or in digging. They eat worms and insect larvae. Captive specimens never bite. They burrow rapidly in moist sand or gravelly soil, remaining close to the moisture line. The two similar species are the only American snakes without large belly scales. Blind Snakes lay eggs. They are relatives of the Boas.

72

RUBBER BOA

ROSY BOA

BOAS are not all large tropical snakes. Two species live north of Mexico. The Rosy Boa, an attractive, docile, small-scaled constrictor, lives in dry, rocky foothills. The grayish Rubber Boa, also heavy-bodied, has a short, blunt tail which it displays like a head while its real head is protected by the coils of its body. It grows up to 2 ft. long; the Rosy Boa is larger (3 ft.). Both bear live young, and are protected through-out their ranges.

Rubber

Rosy

73

RAINBOW SNAKE has stripes that vary from orange to red. The underside is red with a double row of black spots. A snake of swampy regions, it often burrows and is not commonly seen. It swims well and is very good at catching eels, its main food. It is smaller (40 in.) than

the closely related Mud Snake (p. 75), but has a similar sharp "spine" at the end of its tail. The female Rainbow Snake lays 20 or more eggs, which hatch after about 60 days.

MUD SNAKE is the subject of many superstitions. The spike or stinger on the tail is erroneously said to be poisonous. This snake, also called Hoop Snake, is supposed to grasp tail in mouth and roll like a wheel. Such tales about the harmless, attractive, small-headed Mud Snake are untrue. This burrowing swamp snake feeds on fish and frogs, especially on Sirens and Amphiumas (pp. 139-140). Length, 4 to 6 ft.; lays 20 to 80 or more eggs. Two races.

NORTHERN
RINGNECK
SNAKE

Eastern Group of Ringneck Snakes

Western Group of Ringneck Snakes

Variations in Belly Markings

Northern   Southern   Prairie

RINGNECK SNAKE, one species with 12 subspecies in the U.S., is a small (12 to 18 in.), attractive snake living in moist woods under rocks or fallen logs, where it feeds on small insects and worms. It lays eggs hatching in about two months. Recognize this snake by its slate-gray color and usually the yellow-to-orange ring behind the head. The underside is yellow, orange, or red, sometimes spotted. It may secrete a smelly fluid when captured, but does not bite.

ROUGH GREEN SNAKE

eeled Scale

SMOOTH GREEN SNAKE

GREEN SNAKES live in greenery where they are hard to see. They are slender and feed on insects and spiders. The Smooth Green Snake (15 to 18 in.) has smooth scales and prefers open grassy places. The Rough Green Snake grows twice as long. It has a rough appearance because of a ridge or keel on each scale. It is an excellent climber and prefers bushes and vines. The young of both are dark olive or bluish gray, but gradually turn green.

Smooth

Rough

77

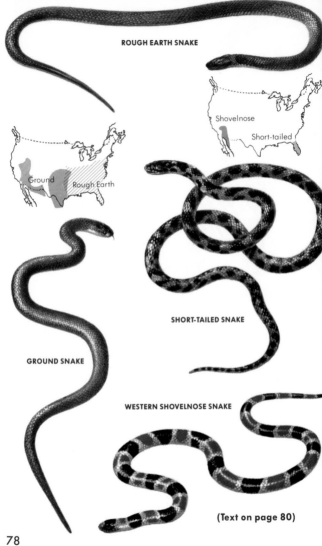

ROUGH EARTH SNAKE

Shovelnose
Short-tailed

Ground
Rough Earth

SHORT-TAILED SNAKE

GROUND SNAKE

WESTERN SHOVELNOSE SNAKE

(Text on page 80)

78

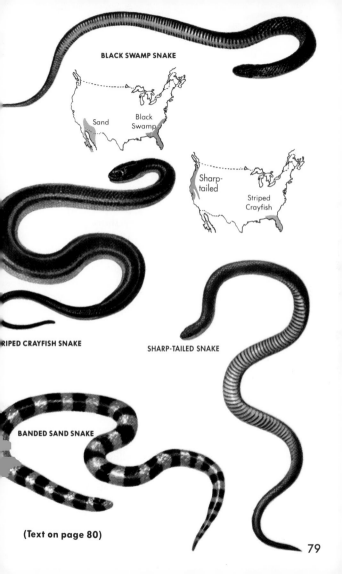

BLACK SWAMP SNAKE

Sand

Black
Swamp

Sharp-
tailed

Striped
Crayfish

STRIPED CRAYFISH SNAKE

SHARP-TAILED SNAKE

BANDED SAND SNAKE

(Text on page 80)

BANDED SAND SNAKE

SMALLER, LESS COMMON, HARMLESS SNAKES

(Illustrations on pp. 78 and 79)

EARTH SNAKES (10 to 12 in.) are two woodland species, one Rough, the other Smooth. Brownish or gray above, same with small black dots. Food: small insects and worms. Young are born alive.

SHORT-TAILED SNAKE (18 to 24 in.) is like a small, slender Eastern Milk Snake (p. 98). An aggressive, burrowing, upland snake, it kills small prey, often other snakes, by constriction. Tail is very short.

GROUND SNAKE (10 to 15 in.) is a small species of exceptionally variable color and pattern, similar but not related to Sharptail. Food: insects, spiders, etc.

SHOVELNOSE SNAKES (12 to 16 in.) are ground snakes (two species) slightly larger than Ground Snake and related to it. Snout projecting but flattened. A yellowish, egg-laying sand burrower.

BLACK SWAMP SNAKE (12 to 16 in.) is thick-bodied, red-bellied, swamp-loving. Black bar on each belly scale. Young born alive. Food: probably fish, frogs.

STRIPED CRAYFISH SNAKE (18 to 24 in.) is aquatic, living in holes and tunnels along ditches and in swamps. Food: mainly crayfish and frogs. Young are born alive.

SHARP-TAILED SNAKE (12 to 16 in.) is somewhat stout. Little is known of its habits. Note the light yellow stripe on sides, black bands on yellow belly scales.

BANDED SAND SNAKE (10 to 14 in.) is a burrower in desert sands. Crawls just below the surface, aided by a broad, heavy snout. Yellow to red, with dark bands almost encircling body. Scales small and shiny. Life history largely unknown. Feeds on burrowing insects.

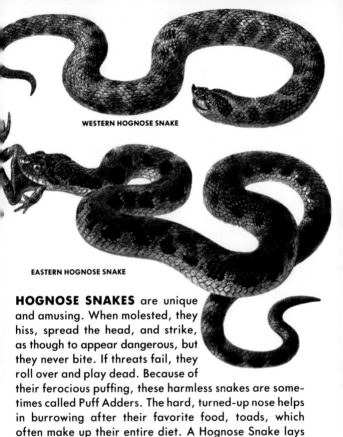

WESTERN HOGNOSE SNAKE

EASTERN HOGNOSE SNAKE

HOGNOSE SNAKES are unique and amusing. When molested, they hiss, spread the head, and strike, as though to appear dangerous, but they never bite. If threats fail, they roll over and play dead. Because of their ferocious puffing, these harmless snakes are some-times called Puff Adders. The hard, turned-up nose helps in burrowing after their favorite food, toads, which often make up their entire diet. A Hognose Snake lays about two dozen eggs in summer. These gentle snakes do well in captivity if fed toads or lizards. Eastern Hognose Snake, 2 to 3 ft. long, is the largest of three similar species, all heavily built. Southern species, like Western, has sharply turned-up nose.

81

BROWN VINE SNAKE

CHIHUAHUAN HOOK-NOSED SNAKE

CHIHUAHUAN HOOK-NOSED SNAKE

NORTHERN CAT-EYED SNAKE

Chihuahuan
Hook-nosed

Cat-eyed

Brown
Vine

Black-
striped

(Text on Page 84)

82

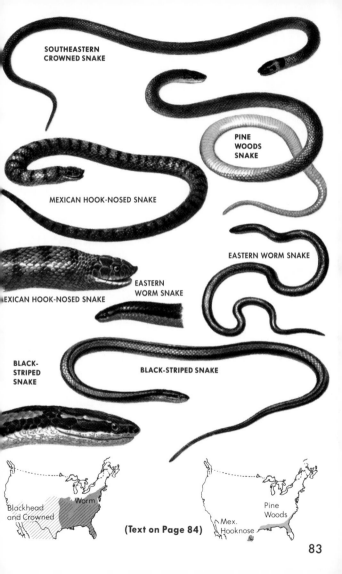

SOUTHEASTERN CROWNED SNAKE

PINE WOODS SNAKE

MEXICAN HOOK-NOSED SNAKE

EASTERN WORM SNAKE

MEXICAN HOOK-NOSED SNAKE

EASTERN WORM SNAKE

BLACK-STRIPED SNAKE

BLACK-STRIPED SNAKE

Blackhead and Crowned

Worm

Pine Woods

Mex. Hooknose

(Text on Page 84)

83

(Illustrations on pp. 82–83)

BROWN VINE SNAKE* (to 4 ft. or more) is a bush-dweller of semiarid regions, very slender. Longer, narrower head than other American snakes. Reddish-brown; white line down belly. Food: lizards.

CHIHUAHUAN HOOK-NOSED SNAKE, a blotched, egg-laying burrower, resembles a miniature (10 to 12 in.) Hognose Snake but is not kin.

CAT-EYED SNAKE (to 30 in.), a wide-headed, slender egg-layer, feeds on both invertebrates and small vertebrates. Often found in trees and bushes.

BLACKHEAD and CROWNED SNAKES* (12 to 14 in.) are a large group of secretive or burrowing, egg-laying species. All but one have a black head cap.

PINE WOODS SNAKE (12 to 16 in.) has a yellow upper lip. Back reddish-brown, belly yellow. An egg-layer of swamps, found under logs and debris. Food: frogs, toads, insects.

MEXICAN HOOK-NOSED SNAKE (10 to 12 in.), related to the Western, is subtropical, with a larger shovel-snout than the Western but with similar habits. Ashy gray with gray and black cross bands.

WORM SNAKE (10 to 13 in.) is a burrower, rarely seen. Shiny, smooth scales. An egg-layer; feeds on earthworms. Found in woods.

BLACK-STRIPED SNAKE* (16 to 20 in.) is a night snake. Eats frogs, toads, lizards. An egg-layer and ground-dweller. Rare; commoner in tropical America.

*Species with weak venom and small, fixed, grooved fangs in rear of upper jaw.

WESTERN YELLOWBELLY RACER

NORTHERN BLACK RACER

RACER (one species, with ten subspecies) is aggressive and graceful. In the East, they average 4 ft., are smooth, blue-black, with white chin and throat. Western form is smaller, greenish or yellowish brown, with belly and chin lighter. All Racers are very active, at home in bushes and trees. Their food is small mammals, birds, insects, frogs, lizards, and other snakes.

YOUNG RACER

Western | Racers | Black

85

COACHWHIP AND WHIPSNAKES are closely related to Racers, but grow longer. Coachwhips range in color from reddish or yellow-brown to dark brown. Most are the same color all over, but some are darker at the head and lighter toward the tail. Coachwhips are the most widely distributed of this group. They are also the largest, with some measuring over 7 ft. long. They move very quickly, roaming across open ground with their head held high above the ground. Whipsnakes grow to 4 or 5 ft. long. They typically have a dark back with

STRIPED WHIPSNAKE

yellow stripes on the sides and a lighter colored belly. They are alert and fast, and active during the day, even in the heat of the desert. Whipsnakes are always aggressive and nervous. They feed on mice, lizards, and small snakes, moving rapidly over sand or through brush after their prey. They do not kill by constriction, but chew when they bite. Eight to 12 eggs are laid in early summer, each about 1 by 1½ in.

Whipsnakes

Coachwhip Snake

MOUNTAIN PATCHNOSE SNAKE

PATCHNOSE SNAKES are as fast and as active as their relatives, the Racers. They hunt lizards, snakes, and small rodents during the day in almost any terrain. A patchnose snake can be identified by the blunt shield over its nose and the yellow and brown stripes down the back. The nose may help when burrowing in sand. Rear teeth are enlarged but venom is weak. Adults are about 3 ft. long. Females lay eggs.

SADDLED LEAFNOSE SNAKE

LEAFNOSE SNAKES, two small relatives of patch-noses, have an even more exaggerated nosepiece. They are secretive and nocturnal, but are seen fairly commonly along highways in deserts at night. The two species are marked by dark blotches. Both are moderately stout, 12 to 15 in. long. They are pugnacious, coiling and striking when caught, but are harmless. They are egg-layers, and are reported to feed on desert lizards or lizard eggs.

89

GRAY RAT SNAKE

RAT SNAKES include five large species, common and widely distributed in the East and Middle West. Colors differ from species to species, making identification easier. All are fast, active snakes. When caught they may bite freely and excrete a foul-smelling fluid from glands at base of the tail. They are not consistently aggressive, but bite unpredictably. All are constrictors. Members of this group have been known by diverse common names which are often misleading. Names used here are truer to the snakes. Black Rat Snake, also known as Pilot Black Snake, may be mistaken for the Black Racer (p.

YELLOW RAT SNAKE

BLACK RAT SNAKE (adult)

85). Black Rat Snake has some scales tipped with white —remains of a pattern of blotches seen more clearly in young of all members of this group. The scales are slightly keeled; those of the Black Racer are not. Gray Rat Snake, a more southern form, has blotches of gray or brown against a lighter background. Yellow Rat Snake (Striped Chicken Snake) averages 4 to 5 ft. long, sometimes reaches 7 ft. It is dull or olive yellow with four black lines down its back. Often found around barns or stables, it is looking for rats more often than for fowl. The Black, Gray, and Yellow Rat Snakes are all subspecies of one species.

BLACK RAT SNAKE (young)

Other Rat Snakes

Common Rat Snake

CORN SNAKE

CORN AND FOX SNAKES are colorful members of
the rat snake group. The Corn Snake earned its name
because it may be found in areas where corn is stored
and mice accumulate. It is also known as the Red Rat
Snake because of the reddish-brown blotches on its skin,
but a western form lacks this red color. Corn Snakes
grow to 4 ft. long. They are harmless but are sometimes
confused with the Copperhead (p. 109) because of their
color pattern. They spend much of their time underground in rodent burrows, but will also climb. The Fox

FOX SNAKE

Snake is like other rat snakes but is somewhat heavier and climbs less often. It averages 3 to 4 ft. long and is colored with brownish blotches on a straw-yellow background. Found in woods and open country. All rat snakes lay eggs, often in rotted logs or stumps. Corn Snakes and Fox Snakes feed almost exclusively on mice and birds. When disturbed they hiss and vibrate their sharp tail, as will other rat snakes.

Fox Snake

Corn Snake and kin

EASTERN INDIGO SNAKE may reach over 8 ft. long. It is shiny, midnight blue, with red blotches on its head and neck. The Texas Indigo Snake has brownish blotches. Both feed on small mammals and other snakes, even rattlesnakes and Cottonmouths. Indigo Snakes are

often found in the burrows of gopher tortoises (p. 27). If handled, they may bite unpredictably and chew tenaciously. They are rare, especially in the East.

94

GLOSSY SNAKE looks somewhat similar to the Bull-snake (p. 96), but it has smooth scales while those of the Bullsnake are keeled. These smooth, shiny scales are responsible for the Glossy Snake's common name. Blotched, spotted, and gray-brown, these snakes are slender, with narrow heads. They are constrictors, feeding on lizards, rodents, and other small animals. They lay eggs and are nocturnal. Adults average 30 to 36 in. long. Seven races.

NORTHERN PINE SNAKE

Pine Snake Eggs

BULLSNAKES and their kin are found from coast to coast. These large, heavy snakes average 5 ft. long and may grow to 7 ft. They are the most common constrictors, widely known as destroyers of rodents. Bullsnakes have a large, vertical nose plate, adapted for burrowing. All hiss very loudly when angered and will strike to defend themselves, often vibrating the tail tip in warning. Most become docile in captivity, but others remain nervous. The Pine Snake is an eastern form of the Bullsnake, named for its favorite habitat—southern pine woods. It is relatively light-colored with large black patches on the back. Food is small rabbits, squirrels, rats, and mice.

NORTHERN PINE SNAKE

BULLSNAKE

The Bullsnake is more common farther west. It is more yellow than a Pine Snake and has a larger number of dark blotches on its skin. The Pacific Coast form, known as the Gopher Snake, is smaller than a Bullsnake and has more blotches. All these snakes play an important role in controlling rodent popula-tions. They often lurk near clumps of vegetation, waiting for their prey. They also enter underground burrows to hunt for pocket gophers and ground squirrels.

Gopher
Bull
Pine

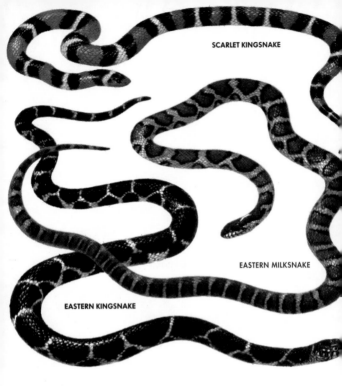

SCARLET KINGSNAKE

EASTERN MILKSNAKE

EASTERN KINGSNAKE

KINGSNAKES AND MILKSNAKES are a group of medium-sized snakes with shiny, smooth scales. They include six species, ranging from southern Canada through much of the U.S. All are constrictors and some

are at least partially immune to the poison of our venomous snakes. Kingsnakes feed on other snakes, but also eat many kinds of rodents. Milk Snake (30 in.) of central and eastern U.S. has red or brown blotches or rings bordered with black; belly is pale, with black patches. It feeds mainly on rodents and not, as fables tell, by milking cows. Eastern Kingsnake is shiny black with bands of yellow crossing in a chainlike pattern. It is larger (3½ to 4 ft.) than a Milk Snake and more common in open country. The small Scarlet Kingsnake (18 in.), like some races of Milk Snake, may be confused with Coral Snakes (p. 108), but note that each yellow band is bordered by black. The Common Kingsnake (3 ft.) is black with white bands; some have a white midline. A speckled form has a light dot on most scales.

SPECKLED KINGSNAKE

COMMON KINGSNAKE

Scarlet Snake

Coral Snake

Scarlet Kingsnake

SCARLET SNAKE (16 to 24 in.) lives underground but is occasionally found under rotting logs or on the ground. It hunts small lizards, snakes, and mice, killing by constriction. Also eats eggs of other snakes. Its markings are similar to the Scarlet Kingsnake (p. 98)—white bands bordered by narrow black bands—but its belly is unmarked. Sometimes mistaken for the venomous Coral Snake (p. 108), which has black bands bordered by yellow and a black snout. Lays eggs.

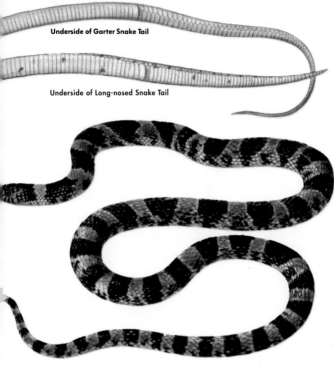

Underside of Garter Snake Tail

Underside of Long-nosed Snake Tail

LONG-NOSED SNAKE (2 to 3 ft.) does not have a long nose, but its small, narrow head gives that appearance. A burrower, it is active from dusk to dawn. It feeds on lizards, snakes, and small mammals killed by constriction. Generally speckled, the color is variable with dark saddles on the back bordered by bands of red, white, or yellow. Belly is lighter. Long-nosed Snake is the only harmless snake in the U.S. with a single row of scales under the tail.

COMMON WATER SNAKE

PLAIN-BELLIED WATER SNAKE

WATER SNAKES are found mainly in the East in almost any fresh water. They show little external adaptation to water life but are fine swimmers and divers. Most seek water when disturbed and to find their food, mainly crayfish, fish, and frogs. All are nonvenomous, but some are easily confused with the venomous southern Cottonmouth (p. 109). Most Water Snakes are badtempered. When caught, they bite and exude a repulsive-smelling fluid mixed with feces from glands at the base of the tail. The Banded Water Snake (30 in.)

GREEN WATER SNAKE

DIAMONDBACK WATER SNAKE

has reddish-brown bands the length of its body. Diamondback Water Snake is larger (3½ to 5 ft.) and darker, with diamond-shaped blotches over the backbone. Plain-bellied Water Snake (3 to 5½ ft.) is dark above, with yellow or reddish belly. Green Water Snake (3 to 5½ ft.) is a dull olive green with a vague, barred pattern. Young Water Snakes are born alive, sometimes in great numbers.

Common

Other

PLAINS GARTER SNAKE

WESTERN TERRESTRIAL GARTER SNAKE and Youn

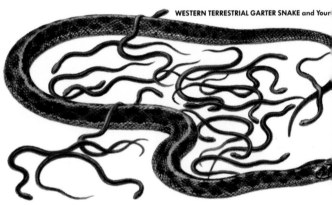

GARTER SNAKES may be more common and better known than any other snake. Most are small (18 to 44 in.), striped, and have keeled scales. Their name comes from their stripes, which look like those on old-fashioned men's garters. Garter Snakes feed on frogs, toads, and earthworms. Young are born alive in summer—often 20 or more at a time. Like Water Snakes they eject unpleasant fluid from glands near their tail when captured.

Other Garter Snakes

Ribbon Snake

Common Garter Snake, more aggressive than others, is marked by three yellowish stripes; the dark area between is spotted. The center stripe of Plains Garter Snake is often a rich orange; the belly is darker than in Common Garter Snake. Several western species have the central stripe brighter than the side ones, and in others the side stripes are brighter than the median one. Ribbon Snakes (two species) are thinner, smaller, with yellow or red stripes against brown scales. Tail is nearly a third of body length.

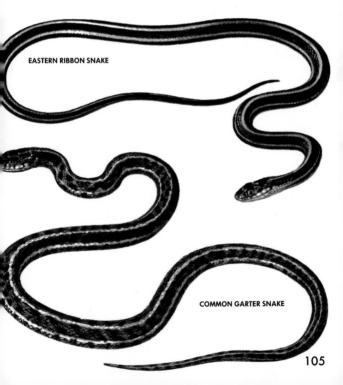

EASTERN RIBBON SNAKE

COMMON GARTER SNAKE

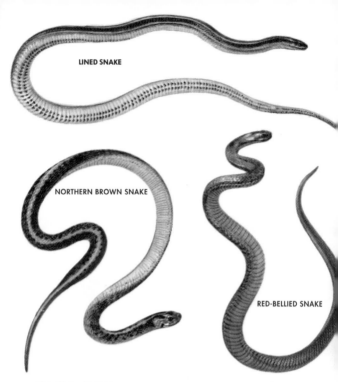

LINED SNAKE

NORTHERN BROWN SNAKE

RED-BELLIED SNAKE

SMALL STRIPED SNAKES are common but inconspicuous. Lined Snake (12 to 20 in.) looks like a miniature Garter Snake with a yellow stripe down its back and black dots on its belly. Northern Brown Snake (10 to 16 in.) is a brownish, burrowing species, common even near cities. Its belly is yellow to pink, with black dots along the sides. Red-bellied Snake (10 to 14 in.) is similar, but has a red belly and yellow spots or collar at the back of its head.

Red-bellied

Lined

Brown

106

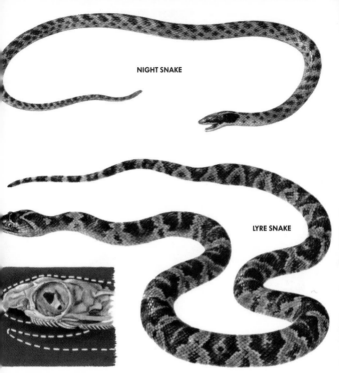

NIGHT SNAKE AND LYRE SNAKE have enlarged teeth in the rear of their jaws, not true fangs. They frequent rocky areas, hunting lizards and small snakes and mammals. When either bites its prey, the venomous saliva immobilizes it. Little is known about how the venom affects humans. Night Snake (15 in.) has two brown blotches on its neck. Lyre Snake (3 ft.) is named for the V or lyre-shape on its head.

107

EASTERN CORAL SNAKE

WESTERN CORAL SNAKE

CORAL SNAKES are highly venomous. Related to cobras, they have a paralytic venom. Red and yellow rings touch on both U.S. Coral Snakes. On the Scarlet Kingsnake (p. 98) and Western Shovelnose (p. 78) red is separated from yellow by black. Eastern Coral Snake

Western   Eastern

(30 to 39 in.) is secretive and burrowing. It is often out in the early morning, hunting mainly lizards and other snakes. The smaller Western Coral Snake (18 in.) has similar habits.

COPPERHEAD

COTTONMOUTH

COPPERHEAD AND COTTONMOUTH are venomous pit vipers, related to rattlers. A pit between the eye and nostril is sensitive to heat. It helps them find warm-blooded prey. Copperheads (30 to 50 in.) are upland snakes with a coppery-red head. Their bodies are marked with coppery-red patches shaped like an hourglass. Cottonmouth (40 to 58 in.) is larger, heavier, and darker, with no strong markings. It is a swamp snake, feeding on fish and frogs.

Copperhead

Cottonmouth

109

| Cross Section of Rattle | Button | Young | Older | Adult | Old Adult |

RATTLESNAKES are often heard before they are seen. Their rattle makes a distinct buzzing sound. They have fangs that pivot outward when their mouth opens. Their venom is lethal and even a dead snake can be danger-ous. They sometimes bite by reflex. Rattlesnakes hunt rabbits, gophers, and other small animals. There are two major groups (genera). Plated rattlesnakes (p. 111) have large scales on the top of their head. Mailed rat-tlesnakes (pp. 112–113) have small scales on theirs. Mailed include the Timber Rattler (3½ to 6 ft.) found in woodlands. It is yellowish with dark, V-shaped bands and a dark tail. The Eastern Diamondback averages 5

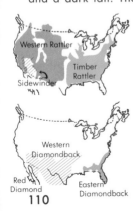

ft., but may grow much longer. The Western Rattler (2½ to 5 ft.) is typ-ically greenish yellow with darker blotches. The Western Diamond-back (4½ to 7½ ft.), found in warm areas, is brown-blotched. The Red Diamondback is similar, but red-dish. The Sidewinder (18 to 30 in.) moves over sand in a rapid, side-wise motion. Young rattlesnakes are born alive; litters of 12 are common.

PIGMY RATTLER

MASSASAUGA

PIGMY RATTLER   TIMBER RATTLER

MASSASAUGA AND PIGMY RATTLER are smaller than their larger relatives, but have larger scales on the top of their head. The Massasauga (2 to 3½ ft.) may not rattle or strike unless provoked. The southern Pigmy Rattler (18 to 24 in.) is ill-tempered. It will often rattle and strike when disturbed, but sometimes its small rattle can barely be heard. It is found in pine and scrub-oak forests.

Massasauga

Pigmy Rattler

111

TIMBER RATTLER
(black phase)

EASTERN DIAMONDBACK RATTLER

WESTERN RATTLER

(Text on Page 110)

112

SIDEWINDER

WESTERN DIAMONDBACK
RATTLER

RED DIAMOND RATTLER

(Text on Page 110)

AMERICAN ALLIGATOR

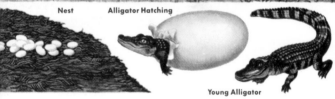

Nest Alligator Hatching

Young Alligator

ALLIGATORS and CROCODILES form a distinct group of reptiles of ancient lineage. Once common in southern swamps, alligators have been reduced in number and range by hunters. Protection in recent years has resulted in recovery in some areas. Large specimens, 10 ft. and over, are still rare. Alligators are not usually dangerous. Reports of "man-eaters" usually refer to crocodiles of Africa or southern Asia. American Crocodile is paler than the alligator; its snout is pointed, narrower. Some of the teeth protrude, bulldog-fashion, from the sides of

AMERICAN CROCODILE

its jaw. Alligators and crocodiles feed on fish, turtles, birds, crayfish, crabs, and other water life. Both lay eggs, hatched by heat of the sun and of decaying vegetation. Crocodiles prefer salt marshes and even swim out into the ocean. Alligators prefer fresh water. Both are protected by law. Tropical American caimans now live in southern Florida. They are small (to 6 ft.) and have a bony ridge between front of orbits.

Alligator

Crocodile

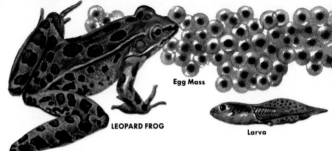

Egg Mass

LEOPARD FROG

Larva

AMPHIBIANS vary considerably in appearance, but most spend at least a part of their lives in water. Eons ago, they were the first animals to venture out of the water to live on land. Those that survive today are still adapted for terrestrial life but most require water for egg-laying and larval development. Unlike reptiles they do not have true clawed feet or scaly skin.

Amphibians are divided into three groups, two of which are common. Frogs and toads are usually tailless when mature and often have well-developed hind legs. The salamanders and their kin are tailed amphibians with a shape similar to lizards. The third group, the Caecilians, are living fossils. They are tropical, burrowing species. Limbless, earless, nearly tailless, and nearly or completely blind, they look like earthworms.

North American amphibians lay jelly-covered eggs one at a time, in clumps, or in strings, usually in quiet water or on moist leaf mold. In most species, these eggs hatch into larvae, or tadpoles, which usually breathe

Lizard Forefoot

Frog Forefoot

Salamander Forefoot

TIGER SALAMANDER

Egg Mass

Larva

with gills and spend much of their life in water. Tadpoles feed mostly on microscopic plants. Eventually they become air-breathing adults which may live in water or on land, returning to water to mate and lay eggs. Adults feed largely on insects. In the North many hibernate underground or in the mud bottoms of ponds during winter.

Frogs and toads are divided into more than two dozen families, eight of which are found in the United States (p. 6). Tailed Frogs (p. 120) represent one family. Another includes the Mexican Burrowing Toad, found only in extreme southern Texas. Other families include Spadefoots (p. 121), True Toads, Treefrogs and Chorus Frogs, Barking and Chirping Frogs (pp. 122–131), and Narrowmouth Toads and True Frogs (pp. 132–136).

The three major groups of salamanders include ten families, eight of which occur in the United States. The terrestrial Slimy Salamanders include by far the largest number of species. These, along with newts and Tiger Salamanders (pp. 144–145), are among the species that spend their adult life on land.

American Toad Calling

1

Extending Its Tongue

2

Catching Fly

FROGS AND TOADS cannot be clearly distinguished, since there are many, not two, distinct groups in their order. True toads usually have rough or warty skins and live mainly on land. Frogs have smoother skins and live in water or wet places. Toads are plump, broad, and less streamlined than frogs. They are slower and cannot jump as well. Some frogs have such varied markings that identification is difficult. Added to this, the skin color and markings of some species change with their surroundings. Most male frogs and toads can inflate a sac in their throat when they make their characteristic sounds. There are eight families of frogs and toads found north of Mexico. The largest are the true toads (pp. 122–123), the treefrogs (pp. 124–129), and the "true" frogs (pp. 132–135).

Skin Color Changes with Surroundings

TADPOLES are the immature or larval stage of frogs and toads. The Barking Frogs (p. 130) are the only native frogs which do not have free-swimming tadpoles. Tadpoles are difficult to identify. The pictures may help you name some species. (See p. 133 for the Bullfrog tadpole.) Collect frogs' eggs or small tadpoles along the shores of ponds and ditches in spring; place them in an aquarium containing pond water and water plants. Do not overstock. As tadpoles hatch and begin to grow they will feed off bits of lettuce, which partly rots in the water. As your tadpoles change into frogs, provide a wooden float on which they can climb and rest.

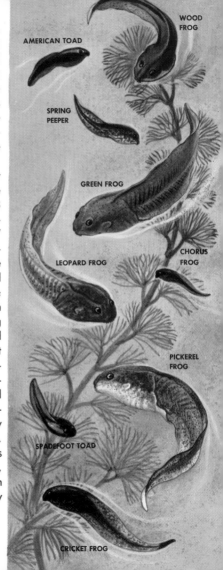

AMERICAN TOAD

WOOD FROG

SPRING PEEPER

GREEN FROG

LEOPARD FROG

CHORUS FROG

PICKEREL FROG

SPADEFOOT TOAD

CRICKET FROG

Female

Male

TAILED FROG is primitive. The male has a distinct tail-like organ. After breeding in late spring or early summer, strings of large eggs are found attached under rocks in rushing, cold mountain streams. Tadpoles cling to the rocks by means of a large sucking disc around the mouth. These small toads, 1 to 2 in. long, vary greatly in color, from gray and black to pink and brown. Note the webbed feet and the short, wide head with a light line sometimes across it.

120

WESTERN SPADEFOOT

EASTERN SPADEFOOT

SPADEFOOTS (six species) have fleshy, webbed feet with large, horny, spadelike warts. In burrowing, the toad corkscrews backward and downward into the soil. It is found under logs or rocks, in shallow holes, coming out at night or after heavy rains to feed. Medium-sized (1½ to 3 in. long), it has relatively smooth skin. Eyes are large, with vertical pupils. Breeding is in late spring and early summer. Eggs are attached to plants at the water's edge.

121

AMERICAN TOAD

TRUE TOADS

TRUE TOADS (18 species in U.S.), a much-maligned group of amphibians, were once wrongly credited with causing warts. Though clumsy, they are well adapted to life on land, feeding on almost any small, moving creature. They protect themselves by burrowing, playing dead, inflating their bodies, and exuding through their skin a white fluid which, in contact with eyes or mouth, is very poisonous. In breeding season and especially when it is raining, males make a species-characteristic trilling call.

American
Woodhouse's

FOWLER'S TOAD

WESTERN TOAD

The American Toad is the common eastern species, 2 to 4 in. long. The male has a darker throat than the female. Fowler's Toad (the eastern race of Woodhouse's Toad) is more greenish and smaller, usually with smaller, more numerous warts and with a white line down the back. The Western Toad (2 to 5 in.) is very warty; the belly is mottled and the head more pointed than in eastern toads. The Great Plains Toad, commonly found along irrigation ditches and streams, is gray or brownish and somewhat varied in pattern.

GREAT PLAINS TOAD

123

CHORUS FROGS are easy to hear but difficult to see. During breeding season they gather near shallow bodies of water and call loudly, often in groups or choruses. In the South this calling begins in November with the first cool rains. In the North it begins soon after the first warm rains of spring. After breeding, females attach small masses of eggs to leaves and stems in water. When the season ends, the frogs seem to disappear entirely, so their habits are not well known. Chorus Frogs have slender bodies, usually less than 2 inches long. They belong to the Treefrog family, but most rarely climb more than a few inches above the ground. Some cannot climb

STRIPED CHORUS FROG

ORNATE CHORUS FROG

at all. The Spring Peeper (p. 126) is the best known in the East. It is light brown with a dark cross on its back. Striped Chorus Frog is brownish or olive-colored, and has distinct stripes on its back. Southern Chorus Frog, common in southern ditches and swamps, has granular olive green skin with irregular spots. Ornate Chorus Frog is chestnut brown, gray, or green with dark spots on its sides and a dark mask. Strecker's Chorus Frog, found farther west, is stockier, usually gray or greenish, with darker blotches.

STRECKER'S CHORUS FROG

SPRING PEEPER

GREEN TREEFROG

SQUIRREL TREEFROG

PINE WOODS TREEFROG

Spring Peeper

Squirrel

(Text on Page 128)

Green

Pine
Woods

126

GRAY TREEFROG

PACIFIC TREEFROG

CANYON TREEFROG

BIRD-VOICED TREEFROG

Canyon   Gray

Pacific

Bird-voiced

(Text on Page 128)

127

TREEFROGS are a large family of small frogs (usually 2 inches or less) that live in trees and shrubs near water. Their calls, heard in early spring, are loud, clear, and musical. True treefrogs, genus Hyla, are lightly built and have sticky pads on their toes that are useful for climbing. Their skin, often warty or rough, varies greatly in color and pattern, making them difficult to tell apart. A few change their color in response to environmental conditions. Spring Peeper, common in woodland swamps, was once considered a treefrog but now is recognized as a Chorus Frog (p. 124). Green Treefrog is slender and long-legged, has smooth green skin, and a penetrating honking call. The smooth skin of a Squirrel Treefrog is green or brown, usually spotted, with a light stripe from eye to forelegs. Pine Woods Treefrog, found in southern forests, ranges from greenish gray to reddish brown, with an irregular dark marking on its back. Its legs are brownish with small orange spots on its thighs. Gray Treefrog is heard in midsummer in woods near water. Its back is spotted or mottled gray or green, thighs orange or yellow. Pacific Treefrog is gray, brown, or green, sometimes spotted. It has a brown V between eyes. Canyon Treefrog can change its color from brown or black to pale pinkish gray. Its skin is rough. The dusky-colored Bird-voiced Treefrog has greenish thighs and dark markings on its back. It utters a unique whistle.

NORTHERN CRICKET FROG

CRICKET FROGS are small Treefrogs (¾ to 1½ in.), but they have no toe pads and cannot climb. Skin is warty; basically brown or green, with a variety of markings in black, red, or yellow. Most have a dark triangle or V-shape on top of their head. Name comes from their call—a sharp, metallic clicking that starts slow and picks up speed. Eggs laid singly, attached to plants in ponds and pools. Two species; common throughout the East.

129

TEXAS BARKING FROG

FREE-TOED FROGS are a mostly tropical American family of many species. The Barking Frog lives in limestone ledges or caves. Eggs are laid in moisture-filled crevices. The tadpoles do not hatch but remain within the egg till they have developed into miniature frogs. The tiny Greenhouse species, imported from the West Indies, is only ⅗ to 1⅕ in. long. The call of the larger Barking Frog (2 to 3½ in.) of Texas sounds like the yapping of a small dog.

BARKING FROG

GREEN-HOUSE FROG

Tadpole in Egg (magnified 3 times)

WHITE-LIPPED FROG

CLIFF CHIRPING FROG

WHITE-LIPPED and CHIRPING FROGS are really Mexican species. The first is a medium-sized, smooth-skinned frog (1½ to 2 in.), marked as its name indicates. It lays eggs in a frothy mass at the edge of ponds. Chirping Frog (three species) is smaller, with more pointed nose and granular skin. Its eggs, laid on land, hatch into legged frogs. There are no free-swimming tadpoles. This dull gray-green frog makes a faint whistling chirp.

Chirping

White-lipped

CRAWFISH FROG

CRAWFISH AND RED-LEGGED FROGS introduce the "true" frog group—21 common species that have smooth, narrow bodies and long hind legs. The Crawfish Frog (2½ to 4½ in.), gray with small black spots, lives in the burrows of Gopher Tortoises or crayfish.

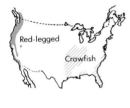

Though fairly common, it is rarely seen. The large Red-legged Frog of the West (2 to 5 in.) is an even dark brown or olive above, colored below as its name indicates. It is a frog of moist forests, breeding in June or July.

RED-LEGGED FROG

GREEN FROG (female)

BULLFROG (male)

BULLFROG and GREEN FROG The Bullfrog is largest of our frogs (4 to 7½ in.). The male has very large "ears" (tympani) behind the eyes; the female's ears are smaller. The color is usually drab green. In the North, the large tadpole does not mature till the second year. The Green Frog is smaller (2 to 4 in.), with a yellowish throat, especially in the males. Both of these common frogs live in ponds and swamps. Both are solitary, laying eggs in spreading surface masses.

Bull   Green

BULLFROG Tadpole

133

PICKEREL FROG   NORTHERN LEOPARD FROG

PICKEREL AND LEOPARD FROGS are commonly found along the shores of permanent lakes or ponds. Pickerel Frogs can be identified by their orange or reddish legs and sides, and square or rectangular spots on their back. Leopard Frogs, a complex of four or more

species once thought to be only one, have greenish legs and sides, and their spots are more rounded. All are slender, smooth-skinned, and about 2 to 4 inches long.

134

WOOD and SPOTTED FROGS

The first is one of the most attractive common frogs: its fawn-brown skin is set off by a dark mask over the eyes. It prefers moist woods, breeds from May to July in woodland pools. Eggs are laid near shore in rounded mass, 2 to 4 in. across, containing 2,000 to 3,000 individual eggs. Length: 1½ to 3 in. The Spotted Frog (3 to 4 in.) is a western species typical of mountain areas. It is dark brown, sometimes spotted with skin slightly roughened. A light streak marks the edge of the upper jaw.

SPOTTED FROG

SHEEP FROG

EASTERN
NARROWMOUTH
TOAD

NARROWMOUTH TOADS have small, wedge-shaped heads with a fold of skin crossing the head just back of the eyes. They are dark or mottled; undersides lighter. Nocturnal toads with tiny voices, they often hide under logs and rocks. The Sheep Frog, a related

species, has narrow head but loose, dark skin, with a narrow yellow or orange stripe down the back. It breeds (March-September) in shallow ponds or large rain-water pools.

Narrowmouth

Sheep

136

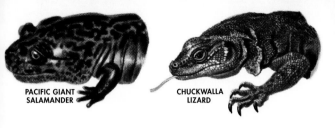

PACIFIC GIANT
SALAMANDER

CHUCKWALLA
LIZARD

SALAMANDERS are tailed amphibians that resemble lizards (pp. 44–45). Unlike lizards, however, they do not have scaly skin or claws, and their front feet never have more than four toes. Lizards usually have five. Salamanders range in size from barely 2 inches to almost 4 feet. Some look very bizarre. All are nocturnal and avoid direct sun. All salamanders need moisture to survive. Some spend their entire lives in water; others live on moist land. Some of those return to water to mate and lay eggs. During breeding season, salamanders move about more and are more likely to be seen. Their eggs are laid singly, in strings, or in small clumps, and have a jelly-like coating.

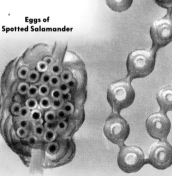

**Eggs of
Hellbender**

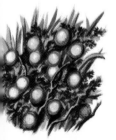

**Eggs of
Dwarf Salamander**

**Eggs of
Spotted Salamander**

MUDPUPPY

MUDPUPPY AND WATERDOGS are large (12 in.) aquatic salamanders of rivers and lakes. Color varies— often dark brown above, paler on belly with dark spots. Larvae throughout life, they have bushy red gills. Eggs are laid in late spring attached to rocks under water. The eggs hatch in 40 to 60 days. Hatchlings, striped on their back and sides, are about an inch long; they mature in about five years. Wrongly thought to be poisonous.

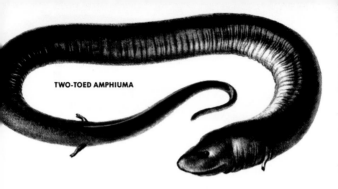

TWO-TOED AMPHIUMA

AMPHIUMAS and HELLBENDER are large aquatic salamanders. The former (three species), smooth and eel-like, grow 30 to 36 in. long, with four tiny, useless, one- to three-toed feet. They are often found in ditches, in burrows, or under debris. The female lays a mass of eggs under mud or rotted leaves. She may remain near to guard them. The Hellbender (16 to 20 in.) is shorter and broader, and lives farther north. Its wrinkled skin makes identification easy. The color varies from spotted yellowish to red and brown. Eggs are laid under rocks in shallow water.

Hellbender
Amphiuma

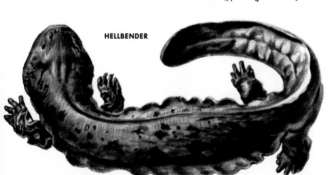

HELLBENDER

GREATER SIREN

SIRENS and DWARF SIREN are southern salamanders of rivers, swamps, and ponds. Both have external gills and both have only front legs. The Sirens (two species) are larger (about 30 in.), gray, olive green, or blackish with spots and blotches. The Dwarf Siren (one species), 5 to 8 in. long, has smaller gills and legs; it occurs in southern streams and waterways. Light stripes down the back and sides are a characteristic marking. Both feed on insects, worms, larvae, and other small water animals.

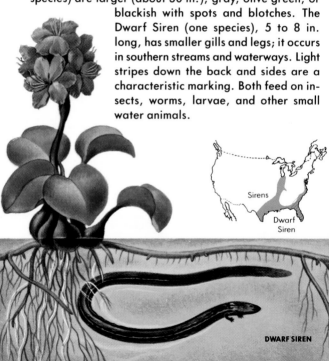

Sirens

Dwarf Siren

DWARF SIREN

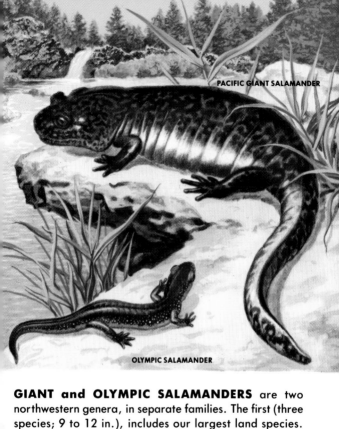

PACIFIC GIANT SALAMANDER

OLYMPIC SALAMANDER

GIANT and OLYMPIC SALAMANDERS are two northwestern genera, in separate families. The first (three species; 9 to 12 in.), includes our largest land species. They are found on moist slopes under rocks and logs. Larvae live in nearby streams. The back color varies—usually mottled; legs darker. Olympic Salamander, smaller (3½ in.), prefers the humid coastal conifer-ous forests, where it is usually found in or along clear streams.

Olympic
Giant (red)

141

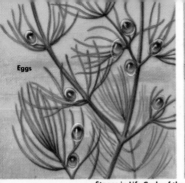

Eggs

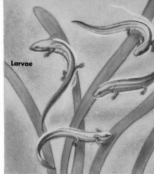

Larvae

Stages in Life Cycle of the Eastern Newts

Red Eft

NEWTS are attractive, interesting salamanders. Of the six species, the eastern (three species, 3 in. long) are perhaps the best known. Their eggs, laid in spring, on stems and leaves of water plants, hatch into larvae. After three or four months in the water these usually leave to spend two or three years on land as an unusual form, known as the Red Eft. When the Efts return permanently to water, they change color and develop a broad swimming tail. Some newts skip the eft stage. Newts feed on worms, insect larvae, and small aquatic

EASTERN (RED-SPOTTED) NEWT—Adults

PACIFIC (CALIFORNIA) NEWT

animals. Red Efts, fed on live insects, do well in terrariums. Adults thrive in aquariums, feeding on small bits of meat. Pacific Newts (three species) are about twice the size of eastern species. They are reddish or dark brown with a lighter yellow or orange belly. Adults are land-dwellers found in moist woods and near mountain pools. They return to water only to breed. Newts secrete a liquid from their skin that predators find bad-tasting.

Pacific

Eastern

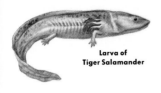
Larva of
Tiger Salamander

SPOTTED SALAMANDER (7 in.) has large, round, yellow or orange spots on a black skin. Like others in this group (16 species) it has vertical grooves on its sides. It is found in moist woods; breeds in ponds and temporary pools. Adults migrate considerably, returning to water to breed. They feed on worms, grubs, and insects.

TIGER SALAMANDER (8 in.) is like the Spotted, but the spots, when present, are larger, more irregular, and extend down the sides and onto the belly. Some larvae do not develop into the land form; they spend their entire life in water, where they eventually breed. Tiger Salamanders are known to live over ten years.

MARBLED SALAMANDER (4 in.) is smaller than others in this group, but like most is a stout, thick-set creature. Variable markings on the black skin, white on males, grayish on females, in irregular fused bands. The larvae are a mottled brown.

Spotted
Tiger
Smallmouth

JEFFERSON SALAMANDER is slender (6½ in.); also called Bluespotted, for the markings on its brownish skin. It lives in woods along swamps and streams. Trunk and tail have vertical grooves.

SMALLMOUTH SALAMANDER (5½ in.) is found in varying habitats from swampy lowlands to upland woods. It is a burrower beneath logs and rocks near streams. The color is a faintly blotched slate gray or brown, lighter beneath.

Jefferson
Related
species
Marbled

SPOTTED SALAMANDER

TIGER SALAMANDER

MARBLED SALAMANDER

JEFFERSON SALAMANDER

SMALLMOUTH SALAMANDER

145

NORTHERN
DUSKY
SALAMANDER

MOUNTAIN
DUSKY
SALAMANDER

DUSKY SALAMANDERS (3½ in.) are very difficult to tell apart. Individuals vary according to their age or size, and even their location. Most have dark, mottled skin that blends with rocks and moss along streams where they live. Many have a light line that runs from

their eye to the corner of their mouth. Their sides have vertical grooves. They are most abundant in the Appalachians.

REDBACK SALAMANDER
(dark phase)

REDBACK SALAMANDERS
(two color phases)

REDBACK SALAMANDER
(red phase)

NORTHERN SLIMY SALAMANDER

REDBACK AND SLIMY SALAMANDERS are land species often found in leaf mold or under logs. Both breed on land and lay eggs in moist nests in rotted bark or logs. Redback (3 in. long) has two color phases: only one has the red stripe down the back. The Slimy Salamander (6 in.) has blue-black skin with small, irregular light spots on its back, and a grayish belly. Its skin is coated with a sticky, glue-like substance.

Redback

Western species

Slimy

147

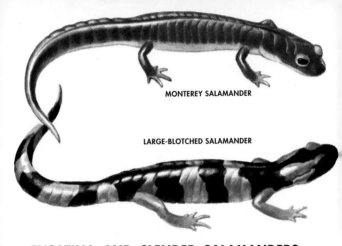

MONTEREY SALAMANDER

LARGE-BLOTCHED SALAMANDER

ENSATINA AND SLENDER SALAMANDERS are western species. The single species of Ensatina (seven subspecies) varies from black to red, usually with red or yellow-orange blotches. These medium-sized salamanders (4 to 5 in.) occur in the mountains, in oak and evergreen forests. They exhibit an unusual, complex courtship pattern. Slender Salamanders (ten species; 4 in.) are thin and wormlike, with a long tail. Color is dark, often spotted or streaked. They are found under rotted logs or leaves where they lay their eggs, from which tiny miniatures of the adults emerge.

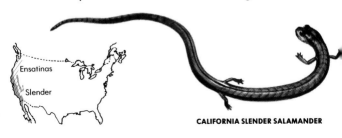

Ensatinas

Slender

CALIFORNIA SLENDER SALAMANDER

ARBOREAL SALAMANDER

CLIMBING SALAMANDERS (five species; 4 in. long) live on both sides of the country. The Green Salamander is found under logs or in crevices on rocky hillsides in the Appalachians. Its skin is dark with greenish blotches. The Arboreal Salamander of the Pacific Coast is often found in water-soaked cavities of trees. Sometimes a whole colony lives in one of these holes, even laying eggs there. They are light brown, with a paler underside and few, if any, markings. Most have long teeth at the front of the upper jaw. Arboreal Salamanders also live on the ground, under logs, bark, and rocks.

GREEN SALAMANDER

Western species

Green

GROTTO SALAMANDER

TEXAS BLIND SALAMANDER

BLIND SALAMANDERS are unusual animals found only in deep wells and underground streams of caves. They are a pale yellowish in color, with eyes reduced in size or completely undeveloped. The larvae of the Grotto Salamander (adults 3¾ in.), found in open streams,

have dark-colored skins and normal eyes. The young of the two Texas species (adults 4 in.) resemble the pale adults. Another rare Blind Salamander has been found in Georgia and Florida.

RED SALAMANDER

SPRING SALAMANDER

RED SALAMANDER
(young)

SPRING and RED SALAMANDERS sometimes prey upon other salamanders. The Spring Salamanders (three species; 5 in. long) are brownish or reddish brown, with vague spots or blotches. Young adults, newly transformed from larvae, are brighter red. This is also true of Red Salamanders (two species; 5 in.). Young adults are bright red with small dark spots; older ones, dull and darker. Both of these salamanders are commonest in hilly or mountain areas along streams or near ponds.

Spring

Red

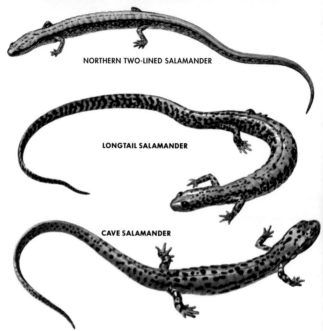

NORTHERN TWO-LINED SALAMANDER

LONGTAIL SALAMANDER

CAVE SALAMANDER

TWO-LINED, LONGTAIL, and CAVE SALAMANDERS represent a common but inconspicuous group (ten species). Two-lined (3 in.) is so marked, with a broken row of dark dots between the lines on its sides. Longtail (about 5 in.) is thin, yellow to orange, with dark tail bars. Both prefer moist sites under logs and rocks, though the Two-lined also prefers brooksides. The Cave Salamander (5 in.) is seen near the entrances of caves and under moist, overhanging rocks. Color is variable, usually yellow or orange with scattered black spots.

Two-lined

Cave

Longtail

FOUR-TOED SALAMANDER is so called because both front and hind feet are four-toed. It is one of the smallest salamanders (2½ in.), fairly common in wooded areas, swamps, and bogs. The dull red-brown back is mottled with darker patches; the belly is lighter, with brown spots. Males are smaller than females and have longer tails. The female lays her eggs in a mossy cavity and stays with them till they hatch, in about two months. The larvae leave the water in six weeks to complete their development on land. They mature in about two years.

153

FOR MORE INFORMATION

A number of excellent books on reptiles and amphibians are available. Two outstanding guides that are designed to follow this book are:

Smith, Hobart M., *Amphibians of North America (Golden Field Guides)*. St. Martin's Press, New York, 1978.

Smith, Hobart M., and Edmund D, Brodie, Jr., *Reptiles of North America (Golden Field Guides)*. St. Martin's Press, New York, 1982.

The following titles offer more detailed information:

Cogger, Harold G., and Richard G. Zweifel, *Encyclopedia of Reptiles and Amphibians. Second edition.* Academic Press, New York, 1998.

Conant, Roger, Joseph T. Collins, and Isabelle Hunt Conant, *A Field Guide to Reptiles & Amphibians: Eastern and Central North America (Peterson Field Guides)*. Houghton Mifflin Co., Boston, 1998.

Ernst, Carl H., and Roger W. Barbour, *Turtles of the World*. Smithsonian Institution Press, Washington, D.C., 1992.

Pough, F. Harvey, et al, *Herpetology*. Prentice-Hall, New York, 1997.

Stebbins, Robert C., *A Field Guide to Western Reptiles and Amphibians (Peterson's Field Guides)*. Houghton Mifflin Co., Boston, 1998.

WEB SITES

The web is full of interesting information about reptiles and amphibians. A good place to start is the Savannah River Ecology Laboratory Herpetology Lab at www.uga.edu/srelherp/. Some keywords to use in searches for these sites include reptiles, snakes, lizards, turtles, amphibians, frogs, salamanders, and herpetology.

MUSEUMS AND ZOOS are good places for finding additional information about reptiles and amphibians.

Atlanta, GA: Zoo Atlanta
Chicago, IL: Field Museum of Natural History; Brookfield Zoo
Columbia, SC: Riverbanks Zoo
Los Angeles, CA: Los Angeles Zoo; Los Angeles County Museum
Milwaukee, WI: Milwaukee County Zoo
New York, NY: American Museum of Natural History; Bronx Zoo
San Antonio, TX: San Antonio Zoo
San Diego, CA: San Diego Zoo; San Diego Wild Animal Park
St. Louis, MO: St. Louis Zoo
Tucson, AZ: Arizona-Sonora Desert Museum
Washington, DC: National Museum of Natural History; National Zoological Park

SCIENTIFIC NAMES

This list of scientific names is included because the common names for reptiles and amphibians sometimes differ from place to place or time to time. The list includes all the reptiles and amphibians illustrated in this book. It follows the standard practice of listing the name of the genus first and the species second. A third name is a subspecies. If the genus name is abbreviated, it is the same as the genus name that precedes it. The page number for each illustration is listed in bold type.

20 Leatherback: Dermochelys coriacea
Hawksbill: Eretmochelys imbricata
21 Loggerhead: Caretta caretta
Green: Chelonia mydas
22 Sternotherus odoratus
23 Eastern: Kinosternon subrubrum subrubrum
Yellow: K. flavescens
24 Chelydra serpentina
25 Macroclemys temminckii
26 Apalone spinifera
27 Gopherus polyphemus
28–29 Trachemys scripta elegans
30 Pseudemys concinna
31 Deirochelys reticularia
32 Chrysemys picta
33 Chrysemys picta
34 Graptemys kohnii
35 Graptemys geographica
36 Emydoidea blandingii
37 Malaclemys terrapin
38 Terrapene carolina carolina
39 Terrapene ornata
40 Clemmys guttata
41 Clemmys marmorata
42 Clemmys muhlenbergii
43 Clemmys insculpta
46 Leaf-toed: Phyllodactylus xanti
Ashy: Sphaerodactylus elegans
Mediterranean: Hemidactylus turcicus
47 Western Banded: Coleonyx variegatus
48 Anole: Anolis carolinensis
Chameleon: Chameleo vulgaris
49 Sauromalus obesus
50 Dipsosaurus dorsalis

51 True: Iguana iguana
Spiny: Ctenosaura pectinata
52 Crotaphytus collaris
53 Gambelia wislizenii
54 Tree: Urosaurus ornatus
Side-blotched: Uta stansburiana
55 Lesser Earless: Holbrookia maculata
Zebra-tailed: Callisaurus draconoides
Colorado Desert Fringed-toed: Uma notata
56 Texas Spiny: Sceloporus olivaceus
Western Fence: S. occidentalis
Sagebrush: S. graciosus
Crevice Spiny: S. poinsettii
Desert Spiny: S. magister
57 Sceloporus undulatus
58 Desert Horned: Phrynosoma platyrhinos
Short-horned: P. douglasii
59 Phrynosoma cornutum
60 Granite Night: Xantusia henshawi
Desert Night: X. vigilis
62 Broadhead: Eumeces laticeps
Western: E. skiltonianus
Gilbert's: E. gilberti
Great Plains: E. obsoletus
63 Eumeces fasciatus
64 Ground: Scincella lateralis
Sand: Neoseps reynoldsi
65 Six-lined Racerunner: Cnemidophorus sexlineatus
Western Whiptail: C. tigris
66 Texas: Gerrhonotus liocephalus infernalis
Southern: Elgaria multicarinata
67 Ophisaurus ventralis

155

68 Florida Worm: Rhineura floridana
California Legless: Anniella pulchra
69 Heloderma suspectum
72 Leptotyphlops humilis
73 Rubber: Charina bottae
Rosy: C. trivirgata
74 Farancia erytrogramma
75 Farancia abacura
76 Diadophis punctatus
77 Rough: Opheodrys aestivus
Smooth: O. vernalis
78 Rough Earth: Virginia striatula
Ground: Sonora semiannulata
Short-tailed: Stilosoma extenuatum
Western Shovelnose: Chionactis occipitalis
79 Black Swamp: Seminatrix pygaea
Striped Crayfish: Regina alleni
Sharp-tailed: Contia tenuis
Banded Sand: Chilomeniscus cinctus
80 Chilomeniscus cinctus
81 Western: Heterodon nasicus
Eastern: H. platirhinos
82 Brown Vine: Oxybelis aeneus
Chihuahuan Hook-Nosed: Gyalopion canum
Northern Cat-eyed: Leptodeira septentrionalis
83 Southeastern Crowned: Tantilla coronata
Pine Woods: Rhadinaea flavilata
Mexican Hooknose: Ficimia streckeri
Eastern Worm: Carphophis amoenus
Black-striped: Coniophanes imperialis
85 Western Yellowbelly: Coluber constrictor mormon
Northern Black: C. constrictor constrictor
86 Masticophis flagellum flagellum
87 Masticophis taeniatus
88 Salvadora grahamiae
89 Phyllorhynchus browni

90 Gray: Elaphe obsoleta spiloides
Yellow: E. obsoleta quadrivittata
91 Elaphe obsoleta obsoleta
92 Elaphe guttata
93 Elaphe vulpina
94 Drymarchon corais couperi
95 Arizona elegans
96 Pituophis melanoleucus
97 Pituophis melanoleucus sayi
98 Scarlet: Lampropeltis triangulum elapsoides
Eastern Milk: L. triangulum triangulum
Eastern: L. getula getula
99 Speckled: Lampropeltis getula holbrooki
California: L. getula californiae
100 Cemophora coccinea
101 Rhinocheilus lecontei
102 Northern Banded: Nerodia sipedon
Plain-bellied: N. erythrogaster
103 Western Green: Nerodia cyclopion
Diamondback: N. rhombifer
104 Plains: Thamnophis radix
Western Terrestrial: T. elegans
105 Eastern Ribbon: Thamnophis sauritus sauritus
Common Garter: T. sirtalis
106 Lined: Tropidoclonion lineatum
Brown: Storeria dekayi
Red-bellied: S. occipitomaculata
107 Night: Hypsiglena torquata
Lyre: Trimorphodon biscutatus
108 Eastern: Micrurus fulvius
Western: Micruroides euryxanthus
109 Copperhead: Agkistrodon contortrix
Cottonmouth: A. piscivorus
111 Pigmy: Sistrurus miliarius
Massasauga: S. catenatus
112 Timber: Crotalus horridus
Eastern Diamondback: C. adamanteus
Western: C. viridis
113 Sidewinder: Crotalus cerastes
Western Diamondback: C. atrox

Red Diamond: C. ruber

114 Alligator mississippiensis

115 Crocodylus acutus

118 Bufo americanus

120 Ascaphus truei

121 Western: Scaphiopus hammondii
Eastern: S. holbrookii

122 American: Bufo americanus
Fowler's: B. woodhousii fowleri

123 Western: Bufo boreas
Great Plains: B.cognatus

124 Southern: Pseudacris nigrita
nigrita
Striped: P. triseriata

125 Ornate: Pseudacris ornata
Strecker's: P. streckeri

126 Spring Peeper: Pseudacris
crucifer
Green: Hyla cinerea
Squirrel: H. squirella
Pine Woods: H. femoralis

127 Gray: Hyla versicolor
Pacific: H. regilla
Canyon: H. arenicolor
Bird-voiced: H. avivoca

129 Acris crepitans

130 Barking: Hylactophryne augusti
Greenhouse: Eleutherodactylus
planirostris

131 White-lipped: Letodactylus
labialis
Cliff Chirping: Syrrhophus
marnockii

132 Crayfish: Rana areolata
Red-legged: R. aurora

133 Green: Rana clamitans melanota
Bullfrog: R. catesbeiana

134 Pickerel: Rana palustris
Northern Leopard: R. pipiens

135 Wood: Rana sylvatica
Spotted: R. pretiosa

136 Great Plains Narrowmouth:
Gastrophryne olivacea
Eastern Narrowmouth: G.
carolinensis
Sheep: Hypopachus variolosus

137 Pacific Giant: Dicamptodon
ensatus
Chuckwalla: Sauromalus obesus

138 Necturus maculosus

139 Amphiuma: Amphiuma means
Hellbender: Cryptobranchus
alleganiensis

140 Greater: Siren lacertina
Dwarf: Pseudobranchus striatus

141 Pacific Giant: Dicamptodon
ensatus
Olympic: Rhyacotriton olympicus

142 Notophthalmus viridescens

143 Eastern (Red-spotted): Notophthalmus viridescens viridescens
Pacific (California): Taricha
torosa

145 Spotted: Ambystoma maculatum
Tiger: A. tigrinum
Marbled: A. opacum
Jefferson: A. jeffersonianum
Smallmouth: A. texanum

146 Northern: Desmognathus fuscus
Mountain: D. ochrophaeus

147 Redback: Plethodon cinereus
Slimy: P. glutinosus (complex)

148 Monterey: Ensatina eschscholtzii
eschscholtzii
Large-blotched: E. eschscholtzii
klauberi
California Slender: Batrachoseps
attenuatus

149 Arboreal: Aneides lugubris
Green: A. aeneus

150 Grotto: Typhlotriton spelaeus
Texas Blind: Typhlomolge
rathbuni

151 Red: Pseudotriton ruber
Spring: Gyrinophilus porphyriti-
cus

152 Two-lined: Eurycea bislineata
Longtail: E. longicauda longi-
cauda
Cave: E. lucifuga

153 Hemidactylium scutatum

An asterisk (*) designates pages that are illustrated.